Charles Burney

An Essay towards a History of the Principal Comets that Have Appeared since the Year 1742

Including a Particular Detail of the Return of the Famous Comet of 1682

Charles Burney

An Essay towards a History of the Principal Comets that Have Appeared since the Year 1742
Including a Particular Detail of the Return of the Famous Comet of 1682

ISBN/EAN: 9783744762427

Printed in Europe, USA, Canada, Australia, Japan

Cover: Foto ©berggeist007 / pixelio.de

A N

E S S A Y

Towards a HISTORY of

C O M E T S.

AN

ESSAY

Towards a HISTORY of the

PRINCIPAL COMETS

THAT HAVE

Appeared ſince the Year 1742.

Including a particular Detail of the Return of the famous Comet of 1682 in 1759, according to the Calculation and Prediction of Dr. HALLEY. Compiled from the Obſervations of the moſt eminent Aſtronomers of this Century.

With REMARKS and REFLECTIONS upon the PRESENT COMET.

To which is prefixed,

A LETTER upon COMETS,

ADDRESSED

To the MARCHIONESS DU CHATELET,

By the late

M. DE MAUPERTUIS.

With a ſhort Account of the LIFE of that celebrated ASTRONOMER and MATHEMATICIAN.

GLASGOW:

Printed for ROBERT URIE.

MDCCLXX.

A

SUMMARY ACCOUNT

OF THE LIFE OF

M. DE MAUPERTUIS.

PETER Louis Morceau de Maupertuis was born at St. Malo, the 28th of September, 1698, and was there privately educated till he arrived at his ſixteenth year, when he was placed under the celebrated profeſſor of philoſophy M. le Blond, in the college of la Marche, at Paris. He ſoon diſcovered a paſſion for mathematical ſtudies, and particularly for geometry. He likewiſe practiſed inſtrumental muſic in his early years with great ſucceſs; but fixed on no profeſſion till he was twenty, when he entered into the army. He firſt ſerved in the grey muſqueteers; but in the year 1720, his father purchaſed him a company of cavalry in the regiment of La Rocheguyon.

He remained but five years in the army, during which time he purſued his mathematical ſtudies with great vigour; and it was ſoon remarked by M. Freret, and other academicians, that nothing but geometry could ſatisfy his active ſoul, and unbounded thirſt for knowlege. In the year 1723 he was received into the royal academy of ſciences, and read his firſt performance, which was a memoir upon the conſtruction and form of muſical inſtruments, November 15, 1724.

During the firſt years of his admiſſion he did not wholly confine his attention to mathematics; he dipt into natural philoſophy, and diſcovered great knowlege and dexterity in obſervations and experiments upon animals.

If the cuſtom of travelling into remote climates, like the ſages of antiquity, in order to be initiated into the learned myſteries of thoſe times, had ſtill ſubſiſted, no one would have conformed to it with greater eagerneſs than M. de Maupertius. His firſt gratification of this paſſion was to viſit the country which had given birth to Newton; and during his reſidence at London he became as zealous an admirer and follower of that philoſopher as any one of his own countrymen.

His next excurſion was to Baſil in Switzerland, where he formed a friendſhip with the famous John Bernouilli and his family, which continued to his death.

At his return to Paris, he applied himſelf to his favourite ſtudies with greater zeal than ever,—and how well he fulfilled the duties of an academician, may be gathered by running over the memoirs of the academy, from the year 1724, to 1736; where it appears that he was neither idle, nor occupied by objects of ſmall importance. The moſt ſublime queſtions in geometry, and the relative ſciences, received from his hands that elegance, clearneſs and preciſion, ſo remarkable in all his writings.

In the year 1736 he was ſent by the king of France to the polar circle, to meaſure a degree in order to aſcertain the figure of the earth, accompanied by Meſſrs. Clairaut, Camus, le Monnier, l'Abbé Outhier, and Celſius, the celebrated profeſſor of aſtronomy at Upſal. This diſtinction rendered him ſo famous, that, at his return, he was admitted a member of almoſt every academy in Europe.

In the year 1740, he had an invitation from the king of Pruſſia to go to Berlin, which was too flattering to be refuſed. His rank among men of letters had not wholly effaced his love for his firſt profeſſion, namely, that of arms. He followed his Pruſſian majeſty into the field; and was a witneſs of the diſpoſitions and operations that preceded the battle of Molwitz; but was deprived of the glory of being preſent, when victory declared in favour of his royal patron, by a ſingular kind of adventure. His horſe, during the heat of the aḉtion, running away with him, he fell into the hands of the enemy, and was at firſt but roughly treated by the Auſtrian ſoldiers, to whom he could not make himſelf known, for want of language; but being carried priſoner to Vienna, he received ſuch honours from their imperial majeſties as were never effaced from his memory.

From Vienna he returned to Berlin; but as the reform of the academy which the king of Pruſſia then meditated, was not yet mature, he went again to Paris, where his affairs called him, and was choſen, in 1742, director of the academy of ſciences.

The comet which appeared this year gave riſe to the following letter: it was addreſſed to the celebrated marchioneſs du Chatelet, whoſe love for the ſciences extended even to the ſtudy of mathematics: and ſhe had M. de Maupertuis for her maſter in geometry, and the Newtonian philoſophy.

In 1743 he was received into the French academy. This was the firſt inſtance of the ſame perſon being a member of both the academies at Paris, at the ſame time.

M. de Maupertius again aſſumed the ſoldier at the ſiege of Fribourg, and was pitched upon, by marſhal Cogny and the count d'Argenſon, to carry the news to the French king of the ſurrender of that citadel.

He returned to Berlin in the year 1744, when a marriage was negociated and brought about by the good offices of the queen mother, between our author and mademoiſelle de Borck, a lady of great beauty and merit, and nearly related to M. de Borck, at that time miniſter of ſtate. This determined M. de Maupertuis to ſettle at Berlin, as he was extremely attached to his new ſpouſe, and regarded this alliance as the moſt fortunate circumſtance of his life.

In the year 1746, he was declared by his Pruſſian majeſty, preſident of the royal academy of ſciences at Berlin, and ſoon after by the ſame prince was honoured with the order of Merit.

However, all theſe accumulated honours and advantages, ſo far from leſſening his ardour for the ſciences, ſeemed to furniſh new allurements to labour and application. Not a day paſſed but he produced ſome new project or eſſay for the advancement of knowlege. Nor did he confine himſelf to mathematical ſtudies only: metaphyſics, chymiſtry, botany, polite literature, all ſhared his attention, and contributed to his fame.

But his conſtitution, though naturally robuſt, ſoon felt the effects of this intemperance, in his philoſophical purſuits. Indeed his health had been confiderably impaired before, by the great fatigues of various kinds in which his active mind had involved him. Though from the amazing hardships he had undergone in his northern expedition, most of his future bodily sufferings may be traced. The intense sharpness of the air could only be supported by means of strong liquors, which helped but to lacerate his lungs and bring on a spitting of blood, which began at least twelve years before he died.

Yet still after his bodily strength was thus impaired, his mind seemed to enjoy the greatest vigour, for the best of his writings were produced, and most sublime ideas developed, during the time of his confinement by sickness, when he was unable to occupy his presideal chair at the academy.

M. de Maupertuis took several journeys to St. Malo, during the last years of his life, for the recovery of his health. And though he always received benefit by breathing his native air, yet still, upon his return to Berlin, his disorder likewise returned with greater violence.—His last journey into France was undertaken in the year 1757, when he was obliged, soon after his arrival there, to quit his favourite retreat at St. Malo, on account of the danger and confusion which that town was thrown into, by the arrival of the Engliſh in its neighbourhood.

From thence he went to Bourdeaux, hoping there to meet with a neutral ſhip to carry him to Hamburgh, in his way back to Berlin; but, being diſappointed in that hope, he went to Toulouſe, where he remained ſeven months. He had then thoughts of going to Italy, in hopes a milder climate would reſtore him to health;—but finding himſelf grow worſe, he rather inclined towards Germany, and went to Neufchatel, where for three months he enjoyed the converſation of lord Marſhal, with whom he had formerly been much connected. At length he arrived at Baſil, October 16, 1758, where he was received by his friend Bernoulli, and his family, with the utmoſt tenderneſs and affection. He at firſt found himſelf much better here, than he had been at Neufchatel; but this amendment was of ſhort duration, for as the winter approached, his diſorder returned, accompanied by new and more alarming ſymptoms.

He languiſhed here many months, during which he was attended by M. de la Condamine, the oldeſt and deareſt of his friends. But his fear of alarming madame de Maupertuis, and expoſing her to the hazards and fatigues of a long journey, made him conceal his danger from her, though he ardently wiſhed to ſee her. However, the truth of his ſituation, at length, reached her at Berlin, and ſhe ſet out with the utmoſt precipitation, (for this lady was a pattern of conjugal affection, as well as of all other virtues,) but ſhe was ſtopped on the road, by an expreſs from Meſſ. Bernouilli and de la Condamine, with the melancholy news, that on July 27, 1759, death had put an end to his ſufferings.

A

LETTER UPON COMETS.

ADDRESSED

To the MARCHIONESS DU CHATELET.

Tu ne quæsieris scire nefas.

YOU desire my opinion, madam, concerning the comet which is at present the general topic of conversation throughout Paris; and your wishes are to me commands. But what can I say to you of this star? Shall I enquire into the influence it may have, or the events it may portend? A different star has decided the events of my life; and upon that my fate solely depends. To comets, therefore, I abandon the fate of kings and empires.

It is not a century since astrology was in vogue both in court and city; astronomers, philosophers, and divines, agreed in thinking comets the causes or forerunners of great events. Some few, indeed, rejected this species of divination by astrological rules: a modern author, celebrated for his piety and aſtronomical learning, believed this kind of curioſity more capable of offending an already incenſed God, than of appeaſing his indignation; and yet he could not refrain from giving us a liſt of all the great events which comets have already preceded, or followed *.

Theſe ſtars, after having been ſo long the terror of the world, are ſuddenly fallen into ſuch diſrepute, that they are no longer held capable of producing any thing but colds. It is not the preſent humour to believe, that bodies ſo remote as comets can have any influence upon ſublunary things; or can be meant as ſigns of what will happen in future. For what relation can there be between theſe ſtars, and what paſſes in the councils or armies of kings?

I ſhall not examine the metaphyſical poſſibility of theſe things, either as to the influence, which neighbouring bodies reciprocally have upon each other, or that which the body has on the mind, of which, however, we cannot doubt, ſince on that frequently depends the happineſs or miſery of our lives.

But as to the influence of comets, our knowlege muſt ariſe either from revelation, reaſon, or experience: and we may venture to ſay, that we

* Riccioli Almageſt. lib. viii. cap. 3 and 5.

have not yet found it in any one of theſe ſources of our intelligence.

The exiſtence of an univerſal connection between every part of nature is very certain, as well in the phyſical, as in the moral world: each event being connected with that which precedes, and to that which follows it, as a link of that chain which forms the order and ſucceſſion of things: if it was not placed as it is, the chain would be different, and appertain to another univerſe.

Comets then undoubtedly conſtitute a part of the great chain of nature; but the ſinging of birds, the ſwarming of bees, the minuteſt atom that floats in air, form likewiſe a part of this wonderful concatenation: and it would be equally reaſonable to conſult them, as comets. In vain have we an idea of this ſcale of beings; we draw no advantage from it to enable us to foreſee events, when their dependencies are ſo remote: our ſafeſt rule is, therefore, to be content with diſcovering events, from thoſe things only whoſe connection is more immediate and manifeſt.

We may compare aſtrologers to adepts, who are for ever labouring to draw gold from materials which only contain its principles and moſt trivial ſeeds; they loſe their time and labour, whilſt the rational chymiſt enriches himſelf by extracting gold from earths and minerals, where it is already formed.

To endeavour at discovering the connections that subsist in nature, is no way inconsistent with prudence; but it is downright folly to push these researches too far; as it is the lot only of superior beings to see the dependance of events, from one end to the other, of the chain which supports them. I shall therefore not entertain you with that kind of influence of comets; nor speak but of things within our conceptions, and for which we can give mathematical demonstration, or physical reasons. Neither shall I enter into a detail of all the strange ideas, which some have entertained concerning the origin, and nature of comets.

Kepler, to whom, in other particulars, astronomy has such great obligations, thought it but reasonable, that as the sea has its whales and monsters, the air should have them likewise. These monsters are comets: and he explained, how the excrement of the air engendered them *by an animal faculty*.

Some have believed that comets were created expressly when they were necessary to announce to man the designs of God; and that angels were their conductors. They added, that this explication ſolved all the difficulties that could be raiſed on this point*.

In ſhort, that all poſſible abſurdities on this ſubject might be exhauſted, there have been thoſe who have denied the exiſtence of comets; and who have taken them for falſe appearances, occaſioned by the reflection, or refraction of light. But they alone can comprehend how this reflection, or refraction could be made, without the exiſtence of a body to occaſion it†.

To Ariſtotle, comets were meteors formed of ſublunary exhalations; and this, we may eaſily ſuppoſe, was the opinion of a herd of philoſophers, who neither believed nor thought but in conformity to his doctrine.

Anterior to this, the ideas of comets were more juſt. The Chaldeans, it is ſaid, knew they were durable ſtars, and a kind of planets; and had even calculated their revolutions. Seneca embraced this opinion, and ſpeaks of comets in a manner ſo conformable to our preſent notion of them, that one might ſay, he predicted what experiment and modern obſervation have ſince ratified; for, after allowing them to be real planets, he adds, 'Ought 'we then to be ſurpriſed, if comets, whoſe ap'pearances are ſo rare, ſeem not yet ſubjected to

*Mæſtinus, Tannerus, Arriaga, &c. † Panætius.

'fixed laws; or that we are yet unable to determine the course of ſtars, whoſe returns are not made till after ſuch great intervals of time? It is not quite fifteen hundred years ſince the Greeks aſcertained the number of ſtars and named them: many nations, at preſent, only know the heavens from what their eyes behold, and cannot account for the diſappearance of the moon at certain periods, nor what is the ſhadow that conceals her from us. It is but lately that we ourſelves have had certain knowlege of theſe things; but the day will come, however, when time, and the diligence of poſterity, ſhall diſcloſe that of which we are now ignorant. One age is not ſufficient to make ſuch great diſcoveries, though our time were entirely devoted to them: what then have we to hope, who make ſuch a miſerable diviſion of it, between ſtudy and vice*?'

I ſhall now, madam, proceed to explain to you what aſtronomy and geometry have taught us of comets: and that which cannot be mathematically demonſtrated, I will endeavour to ſupply by ſuch conjectures as to me appear moſt probable, You will think, perhaps, after having for a long time paid too great reſpect to comets, that we begin all at once to regard them with too much indifference.

* Seneca, Natur. Queſt. lib. 7.

To convey to you an idea of the importance of these ſtars, we muſt begin by ſaying, that they are not inferior in nature to planets, nor even to our earth. Their origin appears as ancient, their magnitude ſurpaſſes that of many planets, and the matter of which they are formed is of equal ſolidity; and they even may, like the earth, have their inhabitants. In ſhort, if the regular planets appear in ſome reſpects to have advantages over comets, theſe have reciprocally advantages over planets. Now as comets compoſe a part of the ſyſtem of the world, I cannot make you perfectly underſtand them, without previouſly giving you a ſketch of that ſyſtem. But in order to facilitate your intelligence of it, I wiſh you had now before you Dr. Halley's chart of the ſolar ſyſtem, with the explanation, in which the orbits, or paths of comets, are marked.

The ſun is an immenſe globe of celeſtial fire, or of matter more like fire than any thing we know of; but notwithſtanding its prodigious ſize, it occupies a point only of that infinite ſpace in which it is placed. We muſt not therefore call the point it occupies, either the centre, or extremity of this ſpace, as that would imply a figure and boundaries. Each fixed ſtar is alſo a ſun which belongs to another ſyſtem.

While our ſun revolves on its own axis in the

ſpace of twenty-five days and a half, the matter of which it is compoſed flies off in all directions, extending its rays to a prodigious diſtance; not only as far as our globe, but an immenſe way farther. This matter, of which light is formed, moves with ſuch aſtoniſhing rapidity, that it employs but half a quarter of an hour in travelling from the ſun to the earth. It is reflected back when it falls on bodies it cannot pierce; and it is by this light we perceive the opake bodies of the planets, which reflect it back again to us; for when the ſun is hidden from us under the other hemiſphere, it permits this feeble luſtre to become perceptible.

There are ſix planets which have no light but that which they receive from the ſun: theſe are Mercury, Venus, the Earth, (which cannot be denied a place among them) Mars, Jupiter, and Saturn. Each of them deſcribes a great orbit round the ſun, and as they are all placed at different diſtances perform their revolutions round him in different times. Mercury, which is the neareſt, finiſhes his courſe in three months: next to the orb of Mercury is that of Venus, whoſe revolution is completed in eight months; then the orbit of the Earth, placed between thoſe of Venus and Mars, is run through in one year, by the planet which we inhabit; Mars employs two years to finiſh his course; Jupiter twelve, and Saturn thirty. One remarkable circumstance in the revolutions which these stars make round the sun, is, that they are all performed the same way; that is, apparently, from east to west; which made a famous sect of philosophers * think, that the planets swam in a great vortex of fluid matter, which, turning round the sun, carried them along with it, and was the cause of their motion.

But besides that the laws which govern the motion of planets, if well examined, is repugnant to such a vortex, you will presently perceive that the motion of comets proves the utter impossibility of it. Many planets in performing their revolutions round the sun, turn at the same time on their own axis: perhaps they all have such a rotation; but we are not assured of any, except the earth, which does it in 24 hours; Mars in 25; Jupiter in 10; and Venus. But though all astronomers agree in allowing to this planet a revolution upon its own axis, of which they are assured by the diversity of faces she presents to us, they are not, however, yet agreed upon the time of this revolurion: some say she performs it in 23 hours, and others in 24 days.

I have not mentioned the moon here, as she is not

* The Cartesians.

a planet of the firſt order; nor does ſhe perform her revolution immediately round the ſun, but round the earth, which at the ſame time carries her with her in the orbit ſhe deſcribes. Such planets as theſe are *ſecondary*, or *ſatellites*: and as the earth has one, ſo Jupiter has four, and Saturn five.

It is but lately that the laws, by which the planets moved round the ſun, have been diſcovered; and theſe laws of their motion, which were diſcovered by the happy Kepler, have aſſiſted the great Newton to inveſtigate their cauſes.

He has demonſtrated, that as the planets move in the manner they do, round the ſun, it is neceſſary that there ſhould be a force which draws them continually towards this ſtar; without which, inſtead of deſcribing curve lines, as they do, each of them would deſcribe right lines, and become more and more diſtant from the ſun, to infinity. He has diſcovered the proportion of that centripetal power which retains the planets in their orbits, and by this means has found the nature of thoſe curves which this power will neceſſarily make them deſcribe.

All theſe curves are reducible to conic ſections; and obſervation ſhews us that all planets deſcribe *ellipſes* round the ſun, which are oval curves formed by cutting a cone in a plane oblique to its axis.

It is demonſtrated by geometry, that the ſun cannot be in the centre of theſe ellipſes; but towards one of their extremities, in a point which is called the *focus*: and this focus is ſo much nearer the extremity of the ellipſes as it is the more excentric. The Sun's place is in this point: and from thence it happens, that, in certain times of their revolution, and in particular parts of their orbits, which are called their *perihelions*, the planets are found neareſt the Sun; and that in others (when they are in their *aphelions*) they are moſt diſtant from it. Of the ſix planets we have mentioned, theſe different diſtances are not very conſiderable, becauſe the ellipſes they deſcribe are not very excentric, deviating but little from circular figures. But the ſame law by which we obſerve them to form theſe ellipſes, permitting them to deſcribe ellipſes of every degree of excentricity; the bounds, which nature ſeems to have preſcribed to theſe orbits of the planets would be matter of aſtoniſhment, did we not find a greater diverſity in thoſe deſcribed by the new ſtars. Comets now fulfil what calculation foreſaw, and what ſeemed deficient in nature. Theſe new planets are always ſubject to the ſame law as the other ſix; but making uſe of the utmoſt liberty which this law permits, deſcribe round the Sun ellipſes very excentric, and of all degrees of elongation.

The Sun, placed in the focus common to every ellipſes nearly circular, which the ſix planets deſcribe, is likewiſe found in the focus of all the other ellipſes which comets deſcribe. The revolutions of theſe laſt round him are regulated by the ſame laws as the revolutions of the others: Their orbits once determined by obſervations, we can calculate their different places in the heavens for the reſt of their courſe; and theſe places anſwer to thoſe where we really find comets, with the ſame exactneſs as planets anſwer to thoſe places in the heavens, in which, according to calculation, they ought to be.

The only difference to be found between theſe new planets, and thoſe firſt known, is, firſt, that their orbits being much more excentric than thoſe of regular planets, and the Sun, on this account, being nearer one of their extremities, the diſtances of comets from the Sun are much more various in the different parts of the orbits they deſcribe. Some few (ſuch as that of 1680) approach this ſtar (the Sun) ſo nearly, that in their perihelion they are not a ſixth part of his diameter diſtant from the Sun, and after this proximity they recede to immenſe diſtances, and to complete their courſes, mount even beyond the regions of Saturn.

From hence we may conclude, that if comets are inhabited by living creatures, they muſt be of a temperament extremely different from ours, to enable them to ſupport all theſe viciſſitudes.

Secondly, Comets employ a much longer time than planets, in finiſhing their revolutions round the Sun. The ſloweſt planet, Saturn, compleats his courſe in 30 years; while the ſwifteſt comet employs 75 years for his: and it is highly probable that the greateſt number are many ages performing their courſes. The length of their orbits, and the ſlowneſs of their revolutions, are reaſons why we have not yet been able to aſcertain the return of comets poſitively; while planets never retire from thoſe regions within reach of mortal view; comets appear only during that little part of their courſe which they deſcribe in the neighbourhood of the Earth: the reſt is performed in the moſt diſtant regions of the heavens, during which time they are loſt to us: and when a comet returns, we have no knowlege of it, but from ſearching anterior records of comets which have appeared after equal periods of time; and by comparing the path of that which appears, with the orbits of thoſe, if we are furniſhed with ſufficient obſervations of them for that purpoſe. It is by ſuch methods that we are encouraged to think the period of the comet which appeared in 1682 to be about 75 years; becaufe we find a comet, which exhibited, in its motion, the fame phenomena, did appear in 1607, in 1531, and in 1456. Now it is more than probable that all thefe comets were the fame; but of this we fhall be more certain, if it appear again in 1757 or 1758 *.

From fimiliar reafons, but from an induction of lefs weight, Dr. Halley conjectures the comets of 1661, and of 1532, to have been the fame, taking 129 years to compleat its revolution round the Sun.

But Aftronomers have advanced ftill farther in their refearches concerning the comet which appeared in 1680, and find fuch a number of appearances after equal intervals of time, that they very reafonable conclude its periodical revolution round the Sun to be 575 years.

But what prevents thefe conjectures from having the weight of certainty, is the want of precifion in thofe obfervations which were made by the ancients. They applied themfelves rather to the events which thefe ftars portended, than accurately to obferve their places in the heavens.

We cannot depend upon any obfervations on comets anterior to Tycho Brahe; and till Newton came, we had not the principles of the theory of

* See the articles 1758 and 1759 in the following Effay.

these ſtars. Time only, and a ſufficient number of obſervations, can perfect this theory. Labour alone will be inſufficient to attain that point of knowlege, at which mankind are permitted to arrive. It is a neceſſary leſſon for them, and they may be well aſſured that neither inceſſant application, nor even the moſt ſublime genius, can obtain it; they muſt wait till a certain epocha of time puts them in poſſeſſion of it.

Although the aſtronomy of comets is ſtill far from perfect, and though we cannot yet minutely calculate their courſe, we ought to be, however, highly ſatisfied with the exactneſs which conſiderable parts of the path which comets take, can be determined; and as they are ſubject to the ſame law which moves the other celeſtial bodies, as ſoon as the comet appears, and has marked its orbit by ſome points in the heavens, where it has been obſerved, we determine its courſe by the theory, and the event has, to our wiſhes, anſwered the calculation, in all thoſe comets which have been carefully obſerved, as far, and as long, as our ſight could follow them.

You will, perhaps, aſk me, why then have we not the full extent of the orbits that comets deſcribe, and the preciſe time of their return? It is not owing to any deficiency in the theory, but to the ſcarcity of obſervations, the imperfections of our inſtruments, and the debility of human ſight.

The much extended ellipſes that comets deſcribe, approach ſo nearly to parabolas, that in the part of their courſe where they are viſible to us, there is not any perceptible difference. It is with theſe ſtars as with veſſels, which we ſee ſet ſail for long voyages: from their ſetting out, we can judge in general, towards what region of the earth they may be bound; but we cannot have a certain knowlege of what particular courſe they will ſteer, unleſs we could ſee them deviate from the track common to many countries *.

Theſe parts of the courſe, which comets deſcribe, when in our view, are common to ellipſes (which are curve lines returning into themſelves) and to parabolas, which extend to infinity; in which caſe there is no hopes of a comet's return: and we calculate their places, as if they really deſcribed theſe laſt mentioned curves; becauſe the points where we find theſe comets are ſenſibly the ſame, and their calculations are infinitely eaſier.

* i. e. Their deviation from the common courſe, if ſeen, would enable us to judge of the whole route. Thus Comets, during the ſhort time they are viſible to us, appear to move in a parabola; but could our ſight extend far enough to enable us to ſee them deviate from that track (and turn inwards) we might then, from the information of our ſenſes, conclude their courſe to be in an (extremely excentrical) ellipſis.

But if our ſight could enable us to purſue comets further, or if we could obſerve them with better inſtruments, we ſhould ſee them abandon the parabolic route to follow the elliptic; and be acquainted with the extent of their ellipſes, and the return of the ſtars which deſcribe them.

We cannot ſuſpect the truth of this theory, if we examine the wonderful harmony, which is found in the obſerved courſes of many comets, and thoſe calculated by Sir Iſaac Newton *. I ſhall not therefore ſpin out this letter with the deſpicable ſyſtems, which different aſtronomers have forged on the motion of comets: the opinions of thoſe who ſuppoſed them meteors were not more ridiculous, and all their ſyſtems are as oppoſite to reaſon, as they are contradictory to experience.

The courſe of comets, once regulated, prevents our regarding them as ſupernatural preſages, or as flambeaux lighted up to menace the earth. But while our ſuperior knowlege of comets, compared with that of the ancients, exempts us from theſe fears; it informs us that they may be the phyſical cauſe of very extraordinary events.

Almoſt all the comets of which we have the beſt obſervations, when they arrive in theſe regions of

* See Tables of the motions of ſeveral comets, in his Principia, b. iii. prop. 41, and 42.

the heavens, have approached much nearer the Sun, than the earth ever does. They almoſt all croſs the orbits of Saturn, Jupiter, Mars, and the Earth: and, according to the calculation of Dr. Halley, the comet of 1680, paſſed ſo near the orb of the earth, that on the 11th of November, it was only half the Sun's diameter diſtant from it.

"But, hitherto, none has threatened the earth with a nearer appulſe than that of 1680; for by calculation I find that, Nov. 11, 1 h. 6 min. p. m. that comet was not above the ſemidiameter of the Sun to the northward of the way of the earth: at which time, had the earth been there, the comet would have had a parallax equal to that of the Moon, as I take it. This is ſpoken to aſtronomers: but what might have been the conſequences of ſo near an appulſe, or of a contact, or laſtly of a ſhock of the celeſtial bodies, (which is by no means impoſſible to come to paſs) I leave to be diſcuſſed by the ſtudious of phyſical matters *."

This great aſtronomer has calculated the orbits of 24 comets, of which we were furniſhed with ſufficient obſervations; and has found that theſe ſtars move in all directions; having nothing in common, but that their orbits are all deſcribed round the Sun.

* Phil. Tranſ. N°. 297.

A great ſect of philoſophers* held, that the celeſtial bodies of our ſyſtem had no other motion than that of being carried away in a vaſt vortex of fluid matter round the Sun. Their opinion was founded upon this: that the motions of the planets are all in the ſame direction as that of the Sun, round its own axis. Though the fact is true in general, yet the planets do not ſo ſtrictly follow this direction as they muſt do, if impelled by the general motion of vortices. They would then all move in the ſame plane, which would be that of the ecliptic, or at leaſt in planes parallel to the ecliptic; but they do neither, which has greatly embarraſſed theſe philoſophers.

A great man † has attempted to account for this obliquity in the courſes of planets, with reſpect to the plane of the ecliptic; and we muſt admire the reſources, and wonderful ſagacity he has diſcovered, in defending the vortices againſt this objection. But comets form an invincible obſtacle to this vortex; they not only deviate from the common road, but move freely in all directions: ſome following the order of the ſigns in a plane, very little inclined to the plane of the ecliptic: others in planes which are almoſt perpendicular to

* Carteſians.

† Mr. John Bernoulli, in his diſſertation concerning the Inclination of the orbits of the planets.

it: and ſome there are, whoſe motion is intirely retrograde, which move in an oppoſite direction to that of the planets, and of theſe pretended vortices; which can only be done by theſe ſtars moving againſt an extremely rapid torrent, without ſuffering the leaſt retardation. But thoſe who believe this poſſible may make the experiment by rowing in a boat againſt the current of a river.

I know ſome aſtronomers have believed that the retrograde motion of comets might not be ſuch in fact, but in appearance; and be in reality direct, as the motions of the planets are in ſome ſituations with reſpect to the earth. This might be credited, if we were permitted to diſpoſe of comets as we ſhould judge convenient on this or that ſide the Sun; and if, thus placed, they were capable of anſwering equally to the neceſſary laws of motion in the heavenly bodies. But this matter being well examined, and calculated, as it has been by Sir Iſaac Newton and Dr. Halley, we find the impoſſibility of placing comets juſt where partiality for vortices would require; and we are reduced to the neceſſity of acknowleging them to be really retrograde. In this variety of motions, it is poſſible that a comet may meet with a planet, in its courſe, nay, with our earth; and, if we ſhould ever be ſo circumſtanced, it is but natural to expect the arrival of ſome terrible accident. The mere approach of theſe bodies to each other would, no queſtion, occaſion great changes in their motions, whether by the attraction they would reciprocally exerciſe on each other, or by ſome fluid compreſſed between them. The leaſt of theſe motions, muſt inevitably change the ſituation of the axis and poles of the earth. That part of the globe which was formerly towards the equator, after ſuch an accident, would be found near the poles; and that which was heretofore near the poles, we ſhould find near the equator.

The approach of a comet might produce conſequences ſtill more fatal; I have not yet mentioned their tails, which have been as productive of abſurd notions, as any thing relative to comets; but the moſt probable opinion is, that they are immenſe torrents of exhalations and vapours forced from them by the ardent heat of the Sun. The ſtrongeſt proof of this is, that the tails of comets are only viſible when pretty near the Sun, that their ſize increaſes in proportion to their approximation to it, and decreaſe, and are diſſipated, when they are diſtant from it.

A comet, accompanied by a tail, might paſs ſo near the earth, as to drown it in the torrent that it draws after it; or in an atmoſphere of the ſame nature of that with which it is ſurrounded. The Comet of 1680, which was ſo near the Sun, ſuſtained a heat 28,000 times greater than that of the earth in ſummer. Sir Iſaac Newton, in the various experiments he has made upon the heat of bodies, having calculated the degree of heat which this comet muſt have acquired, found it would be 2,000 times hotter than red hot iron; and that a globe of red hot iron, of the ſize of the earth, would be 50,000 years in cooling. What then muſt we think of the heat which yet remained in that comet, when, returning from the Sun, it croſſed the orbit of the earth? Had it paſſed nearer, the earth muſt have been reduced to aſhes, or vitrified; or, if the tail only had reached us, the earth would have been drowned in liquid flame; in the ſame manner as we ſee a kingdom of ants periſh by the boiling liquor which the labourer pours upon them.

An ingenious author has made ſome very ſingular and daring enquiries, concerning that comet which was expected to burn the earth *: tracing it back from 1680, (the time when it laſt appeared) he finds a comet in 1106, another in 531 or 532. and one at the death of Julius Caeſar, forty-four years before our Saviour: this comet is, with great probability, ſuppoſed to be the ſame, and performs its revolutions in 575 years; and the ſe-

* A new Theory of the earth by Whiſton.

venth period back from 1680, correſponds with the year of the univerſal *deluge*. After what has been ſaid, we may eaſily conjecture in what manner the author may explain every circumſtance of this great event. The comet in its way to the Sun, paſſing near the earth, drowned it with its tail and atmoſphere, which had not then acquired the degree of heat which we have juſt mentioned, and cauſed the rain of forty days which is mentioned in the hiſtory of the deluge. But Whiſton, from the approach of this comet, drew ſtill another conſequence, which exactly agrees with the manner in which the divine writings have taught us to believe the deluge happened. The attraction which the comet and earth reciprocally exerciſe on each other, changed the figure of the latter; and drawing it towards the comet, let out the ſubterraneous waters, by breaking open the fountains of the great abyſs. The ſame author has not only attempted in this manner to explain the deluge, but believes that, ſome time or other, a comet, moſt probably the ſame, will, as it returns from the Sun, ſhed burning and mortal exhalations on us, and occaſion to the inhabitants of the earth all the misfortunes which are predicted to it, at the end of the world; and cauſe, in ſhort, the *conflagration* which is to conſume this unfortunate planet.

However bold his thoughts may be, they at leaſt contain nothing contrary either to reaſon, revelation, or morality. God made uſe of the deluge to exterminate a race of men, whoſe crimes merited chaſtiſement; and will probably, one day or other, in a ſtill more terrible manner, deſtroy without exception all human kind; but he may have conſigned the effects of his anger to phyſical cauſes; for he, who is the great Creator and Mover of each celeſtial body in the univerſe, may have regulated the courſes of them all in ſuch a manner, as to produce theſe great events in the fullneſs of time.

Though you ſhould not be convinced, madam, that the deluge and conflagration of the earth depend upon the comet, at leaſt you will confeſs that its appulſe may occaſion ſome ſuch accident. Gregory, one of the greateſt aſtronomers of the age, has ſpoken in ſuch a manner of comets, as would re eſtabliſh them in all their former terrors. This great man, who has contributed ſo much to the perfecting the theory of theſe ſtars, in one of the corollaries of his excellent work, ſays, 'Hence, 'alſo it follows, that if the tail of a comet ſhould 'touch the atmoſphere of our earth, (or if a part 'of this matter ſcattered and diffuſed about the 'heavens ſhould fall into it) the exhalations of it 'mixed with our atmoſphere, (one fluid with a-

'nother) may cause very sensible changes in our 'air, especially in the animals and vegetables: for 'vapours, as they call them, brought from strange 'and distant regions, and excited by a very in-'tense heat, may be very prejudicial to the inha-'bitants, or products of the earth; wherefore 'those things, which have been observed by all 'nations, and in all ages, to follow the appari-'tions of comets, may happen: and it is a thing 'unworthy of a philosopher, to look upon them 'as false and ridiculous *.'

A comet passing near the earth might so alter its motion, as even to change it into a comet; and instead of continuing its present course in an uniform mild climate of a temperature adapted to man, and the different animals which inhabit it, the earth would be exposed to the greatest vicissitudes, scorched in its perihelion, or frozen by the cold of the utermost regions of the heavens: proceeding thus from one evil to another, till perhaps another comet again changes its course, and reinstates it in its original uniformity.

Another misfortune might still possibly befal the planet which we inhabit; for if a great comet should advance too near the earth, it might force it from its orbit, and oblige it to perform its fu-

* Gregory, Astron. Physic. lib. v. corol. ii. prop. 4.

ture revolutions round the comet; wholly ſubjecting it either by its attraction, or, if I dare uſe the word, by involving it in its vortex: the earth, thus become a ſatellite to the comet, would be carried by it into thoſe diſtant regions which it viſits. Wretched condition for a free born planet, which has ſo long enjoyed a temperate ſky! In ſhort, the comet might in like manner rob us of our moon; and were we to eſcape upon ſuch eaſy terms, we might think ourſelves very well off. But the moſt violent accident that could befal us, would be the percuſſion, or ſhock of a comet againſt our globe, which might break both itſelf and the earth into a thouſand pieces. In ſuch caſe, doubtleſs, both theſe bodies would be deſtroyed; and gravity would immediately form one or many planets out of them. If the earth has never yet undergone theſe cataſtrophes, it has doubtleſs experienced many great changes. The prints of fiſhes, and even petrified fiſhes, which we find in places very diſtant from the ſea on the very ſummit of the higheſt mountains, are inconteſtable medals of ſome one of theſe events.

A leſs violent ſhock, ſuch as would not entirely break our planet, would certainly cauſe great changes in the ſituation of lands and ſeas; the waters during ſuch an accident would be greatly raiſed in ſome parts, and would drown vaſt regions of the ſurface of the earth, from which they would afterwards ſubſide. To ſuch a ſhock as this Dr. Halley attributes the deluge: the irregular diſpoſition of the beds of different kinds of matter which compoſes the earth; the enormous mountains, much more reſembling the ruins of an ancient world, than one in its primitive ſtate; all concur in giving weight to his opinion. This philoſopher conjectures, that the exceſſive cold we find in the north-weſt parts of America, which is ſo little proportioned to the preſent latitude of thoſe places, is the remains of cold in thoſe countries which were formerly ſituated nearer the poles; and that the prodigious quantities of ice which we at preſent find there, is the remains of that by which they were formerly covered, and which is not yet entirely melted.

It is evident that, whatever may befal the earth, the other planets are equally liable to; unleſs the enormous ſize of Jupiter and Saturn ſhould be their protection from the inſults of comets. It would be a very curious ſight for us to behold a comet approach, and fall upon Mars, Venus, or Mercury; and either break it into pieces in our view, or violently carry it away, in order to make it a ſatellite.

Comets may even attempt the ſun himſelf; and though they have not ſufficient power to draw him after them, yet their magnitude and near approach might enable them to remove him from his present place. But Newton secures us from such a removal, by a conjecture drawn from the known analogy between comets and planets. Amongst these last, the least are situate nearest the Sun, and the greatest are most distant.

Newton conjectures that it is the same with comets; that the least only approach very near this star, and the great ones are kept at the greatest distances*; lest, says he, they should disturb the Sun by their attraction. But is it necessary for the system, that the Sun should not be disturbed? Ought he alone to enjoy this prerogative? Or, rather, is it one? If we consider the celestial bodies as material masses only, is their immobility a perfection? is not their motion, at least, as desirable as rest? Or should we regard them as capable of sentiment, is it unfortunate for me that another has the ascendancy? Is not the fate of him who is attracted as desirable as his who attracts?

You will allow, madam, that comets are not so devoid of importance as they at present are generally believed. Every thing evinces them able to produce most destructive changes, both to our earth, and to the whole solar system; from the

* Philosoph. Nat. Princip. Mathemat. lib. iii. prop. 41.

dread of which habit only ſecures us. But after all, we have great reaſon to think ourſelves ſafe from ſuch calamities as theſe juſt mentioned; for the earth being but a point in the immenſity of ſpace, and our lives of ſo very tranſient duration, joined to our certain knowlege that in ſo many thouſand years no accident of this kind has ever happened to the earth; all theſe conſiderations are ſufficient to prevent our apprehenſions of being either witneſſes, or victims, to any ſuch in future. Thunder, however terrible, is but little to be feared by each individual, by reaſon of the ſmall ſpot he engroſſes in that ſpace where the thunder may fall. It is the ſame with the little point we fill in the vaſt duration of time in which theſe important events happen. But though theſe conſiderations annihilate the danger to us, yet they cannot alter its nature.

But ſtill the conſideration that a common misfortune is ſcarcely any misfortune at all, ought to baniſh our fears; for the mortal who ſhould unfortunately be of a conſtitution too robuſt, and who ſhould ſolely ſurvive the accident which had deſtroyed the whole human race, would have the greateſt cauſe of complaint. King of the whole earth, poſſeſſor of all its treaſures, he would languiſh with grief and ſorrow; and his whole life would not be worth the laſt moment of him who expired with what he loved. But I fear I have railed too much at comets; though I cannot reproach myſelf with injuſtice towards them, as they are really capable of occaſioning every cataſtrophe to us that I have mentioned: however, I will make them all the reparation I can, by pointing out to you the advantages they may poſſibly procure us; though I very much doubt your being ſo ſenſibly touched by the hopes of theſe advantages, as alarmed by the fear of changing a ſtate in which you have hitherto ſpent your time ſo agreeably. Our earth has kept the ſituation it now holds in the heavens theſe five or ſix thouſand years, during which time its ſeaſons have been the ſame, and climates diſtributed as we now ſee, and with which we ought to be very well contented, without aſpiring at a milder ſky, or perpetual ſpring.

Nothing, however, would be more eaſy than for a comet to procure us theſe felicities. Its approach, which, as we have ſhown, might cauſe ſo many diſorders, might alſo render our condition much better. 1ſt. By occaſioning a little change in the ſituation of the earth; it might elevate its axis, and fix our ſeaſons to a continual ſpring. 2dly, A ſlight remove of the earth in the orbit ſhe deſcribes round the Sun, would make her deſcribe one more circular, in which ſhe would always be equi-diſtant from the ſtar from which ſhe receives both heat and light. And, 3dly, though it was obſerved above that a comet might raviſh our Moon from us, yet on the other hand, a ſmall comet might be reduced to a lunar ſtate itſelf, and be both condemned to make its revolutions round us, and to illuminate our nights. It is very poſſible that our Moon was originally a little comet, which advancing too near the earth, was taken captive by it, and reduced to a ſtate of ſervitude. Jupiter and Saturn, whoſe magnitudes are much greater than that of the earth, and whoſe power extends further, and over larger comets, can make ſuch acquiſitions with greater facility than the earth: hence perhaps it is, that Jupiter has four moons to attend him, and Saturn five; and though we have ſeen above how dangerous the ſhock of a comet might prove, yet the miſchief may be ſo inconſiderable, as to be fatal only to that part of the globe which receives the blow; we might perhaps eſcape with the cruſh only of one kingdom, while the reſt of the earth would enjoy thoſe rare productions with which ſo great a traveller would preſent them. We ſhould be greatly ſurpriſed, perhaps, to find the fragments of theſe ſo much deſpiſed maſſes compoſed of gold and diamonds: but it is hard to ſay whoſe amazement would be the greater, at our firſt meeting; ours, or thoſe inhabitants of the comet which the ſhock might throw upon the earth ? What figures would each party think the other ?

But there is ſtill another means of enriching ourſelves from the ſpoils of comets. In the diſcourſe concerning the figure of the ſtars, it has been explained how a planet might appropriate to itſelf the tail of a comet; and, without either drowning its inhabitants, or poiſoning them by noiſome vapours, might form it into a ſort of ring or vault, ſuſpended round it on every ſide. It has been ſhown, that the tail of a comet might be ſo circumſtanced, as that the laws of attraction would oblige it in this manner to ſurround the earth: the form of theſe rings has been determined, and the whole anſwers ſo well to what is obſerved round Saturn, that it ſeems rather difficult for a more probable, or natural account of this phaenomenon to be diſcovered; we muſt not therefore be aſtoniſhed, if one day or other we ſhould ſee a like appearance round the earth.

Newton, in his reflections on the courſes of comets through all the regions of the heavens, and the prodigious quantity of vapours they drag after them, confers upon them an employment which is not perhaps the moſt honourable of any in the ſyſtem: he ſuppoſes that they carry water and humidity to repair the loſſes of the other celeſtial bodies. Such a ſupply may perhaps be needful to the planets, but it would infallibly be fatal to their inhabitants, as theſe foreign fluids would differ too much from our own not to be noxious. They doubtleſs would infect both air and water; and the greateſt number of inhabitants would periſh. But nature often ſacrifices ſmall objects to the general good of the univerſe.

Another uſe of comets may be, to repair the loſſes which the Sun ſuſtains, by the continual emiſſion of the matter of which it is formed. When a comet paſſes ſo near it as to penetrate the very atmoſphere which ſurrounds it, this atmoſphere being an obſtacle to its motion, and a check to its velocity, muſt conſequently alter the figure of its orbit, and contract the diſtance of its perihelion. And this diſtance conſtantly diminiſhing at each return of the comet, it muſt, after a certain number of revolutions, fall into that immenſe fire (the Sun) to which it will ſerve for new aliment: for thoſe vapours and atmoſpheres which would drown planets, are incapable of extinguiſhing the Sun.

What theſe comets perform which move round our Sun, thoſe may likewiſe perform which move round other ſuns, (the fixed ſtars) and may relumine ſtars which are nearly extinguiſhed. But this is one of the leaſt benefits we may derive from comets. And now, madam, I have nearly communicated to you my whole cometary knowlege. A time will come, perhaps, when we ſhall all be wiſer upon the ſubject: for Newton's Theory, which has enabled us to determine the orbit of comets, will in time conduct us to the exact period of their revolutions.

But it will not be amiſs to inform you, that although theſe ſtars, while they are viſible to us, are governed by the ſame laws as the other planets, and are like them ſubject to calculation, yet we cannot be aſſured of ſeeing them return at the exact time aſſigned them, or in exactly the ſame orbits. The many accidents they may encounter in their way, ſuch as paſſing by the atmoſphere of the Sun, or meeting with planets, or other comets, may ſo alter their courſes, that after a few revolutions, we ſhall no longer be able to know them again.

I have ſpoken of almoſt every comet, except that which appears at preſent; for this plain reaſon, becauſe I have but little to ſay about it. This comet, which makes ſo much noiſe, is one of the moſt contemptible that ever appeared. Some have been ſeen whoſe apparent magnitude equalled that of the Sun; many whoſe diameters appeared a fourth or fifth part of his diameter. Others, again, have ſhown with various and vivid colours; ſome have appeared of a frightful red, many of a gold colour, and others inveloped in thick ſmoke; and ſome few have been ſaid to have diffuſed a ſulphurous ſmell even down to the very earth. The greateſt number drew tails of an enormous length after them; and the comet of 1680 had one which filled near half the heavens.

This comet appears only as a ſtar of the third or fourth magnitude, and has a tail of between four and five degrees. It was diſcovered by Mr. Grant, the 2d of March, at the foot of Antinous.

If you wiſh for more ſcience in this ſubject, with a ſucceſſion of obſervations, made with the moſt finiſhed accuracy, you will meet with them in that excellent work of M. le Monnier, called the Theory of Comets.

In the mean time, you will, I hope, be ſatisfied with knowing that this comet has paſſed from Antinous into the Swan, and from the Swan into Cepheus, with ſuch rapidity, that it ſometimes ran ſix degrees in twenty-four hours. It proceeds towards the pole, and is not above ten degrees from it. But it abates of its ſpeed; and its light and tail are ſo diminiſhed, that we perceive it moves from the earth; and that for this time, we have nothing either to hope, or to fear from it.

Paris, March 26,
1742.

AN

ESSAY

TOWARDS A HISTORY OF

THE PRINCIPAL COMETS

THAT HAVE

Appeared ſince the Year 1742.

ADVERTISEMENT.

THE following eſſay is intended for the uſe of ſuch only as give mathematicians credit for their calculations of the orbits, or paths of comets in the heavens; and who, (taking it for granted that theſe calculations are juſt) wiſh to gratify their curioſity, as well with regard to the reſult of mathematical demonſtrations, as to the phaenomena, and moſt intereſting particulars of theſe erratic ſtars.

An aſtronomer, armed at all points with *Theorems*, *Problems*, *Corollaries*, *Lemmas*, and *Scholia*, is a very formidable being, and equally inacceſſible to the generality of mankind, with the ſtars about which he writes. And yet, without theſe terrible arms to defend him from ignorance and preſumption, his ſcience would degenerate into judicial aſtrology, and he would be little better than a juggler or a fortune-teller. It is therefore intended in ſketching out the following little hiſtory of comets, to ſave the reader the expence of purchaſing, and trouble of peruſing a great number of difficult and dry treatiſes, and to give him the ſum and ſubſtance of ſuch diſcoveries and concluſions, as have proceeded from the moſt laborious and operoſe calculations, to which human intelligence can reach. And it ſhould be remembered, that tho' the foregoing letter is written in a familiar and ſportive ſtile, and accommodated to the perception of the ladies, and ſuch as are unſkilled in mathematics, yet it is founded on true ſcience: the author having been among the firſt, and moſt able geometricians, who adopted and explained the Newtonian philoſophy in France, and who, by his journey to the polar circle, in order to meaſure a degree, has proved and illuſtrated it, by aſcertaining the figure of the earth to be juſt what our great countryman had always ſuggeſted it to be.

THE HISTORY OF THE PRINCIPAL COMETS

That have appeared ſince the Year 1742.

IN the preceding letter it has been remarked, that Dr. Halley upon the Newtonian theory, had determined the elements of twenty-four comets*. At preſent the number of thoſe that have been accurately obſerved, and whoſe orbits are calculated, is more than doubled. A particular detail of ſuch as are moſt intereſting may therefore be acceptable to our readers. We ſhall, however, only juſt mention the comets of 1702, 1706,

* The elements of a comet are the five articles which determine the poſition and magnitude of the parabola it deſcribes, and which conſtitute its theory; namely, its node, inclination, place of its perihelion, perihelion diſtance, which is the ſquare of the parameter, and the time when the comet arrives at its perihelion.

and 1718, whoſe elements different aſtronomers have determined by the Newtonian method; but ſhall be a little more particular about that of 1729. A comet rendered very ſingular, if not by its brilliancy, at leaſt by other circumſtances. It was firſt perceived at Niſmes, by father Sarabel, a Jeſuit, July 31, between Canis Minor and the Dolphin; it was ſo ſmall and dull, that during moon light, it was ſcarce viſible; however, this father informed Caſſini and the academicians of it, who obſerved it from the 1ſt of Auguſt till the 21ſt of January 1730, when it diſappeared. Their obſervations are publiſhed in the memoirs of the academy of ſciences for the year 1730, and after them Maraldi, in 1742, has calculated the parabolic trajectory, which it deſcribed. Many other aſtronomers have done the like, as the Abbé de la Caille, M. de Liſle, Mr. Kies, aſtronomer at Berlin, M. Struick, &c. It paſſed between the orbit of Mars and that of Jupiter, but much nearer the latter; hence it was always ſo ſmall and moved ſo ſlow, (for it hardly advanced an eighteenth of a degree in the ſix months it was obſerved); at firſt its motion was direct, and then retrograde, like the ſuperior planets. Calculation agreed ſo well with theſe obſervations, that though they amounted to fifty, the difference in longitude exceeds not three minutes, and in latitude only a few ſeconds.

We ſhall paſs over the comet of 1737 calculated in the Philoſophical Tranſactions, N° 446: that of 1739, of which ſeveral aſtronomers have given the elements; and come to that of

1742. This comet, notwithſtanding M. de Maupertuis thinks it ſo contemptible, was obſerved by Mr. Betts at Oxford, who ſuppoſed its magnitude to be, at leaſt, equal to that of the earth.

1743. Two comets appeared this year. The firſt was obſerved by M. Struick, the ſecond by Mr. Klenkenberg; but both were ſmall and offered no remarkable phenomena to common obſervers.

1744. The comet which appeared this year, was firſt ſeen in England, at the obſervatory of the Earl of Macclesfield, December 23, 1743. It ſeems to have been accurately obſerved at Oxford by the reverend Mr. Betts, who, in his journal, January 23, 1744, ſays, 'The comet this evening appeared extremely bright and diſtinct, and the diameter of its nucleus, nearly equal to that of Jupiter, its tail extending above ſixteen degrees from its body;' and adds, 'That on February 23, the prodigious brightneſs it acquired, by its near approach to the ſun, made it viſible in the day time.' The nodes of this comet, and the planet Mercury, were situated within less than half a degree of each other; which gave rise to a report that the comet carried Mercury from its orbit: but, says Mr. Betts, 'Upon computing 'their heliocentric conjunction, which happened 'February 18, I found the comet was, at that 'time, distant from Mercury nearly one third part 'of the great circle; being twice as near the Sun, 'as the planet Mercury.' This was the most considerable comet that had appeared since the year 1680.

M. des Chezeaux (Essais de Physique 1751) observes, that at its first appearance, it had no tail, at least perceptible to the naked eye; but in approaching the Sun, it acquired one which increased every day till it arrived at its perihelion; so that February the 17th, it was forty degrees long, and it still augmented considerably after the perihelion; for though the body of the comet could no longer be seen, the tail was visible two hours before sun rise, 20 or 30 degrees above the horizon, while the body was below it. According to this author, the tail was divided into five large streams, or bands, and must have afforded a strange spectacle, if the earth had been at that time in a favourable position for observing it. This comet, and that of 1742, gave rise to several learned and ingenious works. Soon after M. de Maupertuis' letter, came out the theory of comets by M. le Monnier, 8vo. in which, besides the translation of Halley's Synopsis, is included an introduction and historical supplement concerning the progress of this theory, before and since Newton's time; together with divers interesting particulars relative to the catalogue of the fixed stars, and theory of the Sun: the treatise of M. des Chezeaux, mentioned above: Osservazioni intorno la cometa dell anno 1744, or observations concerning the comet of 1744 by Zanotti, professor at Bologna; and an excellent treatise by the celebrated Euler, called *Theoria Motuum Planetarum et Cometarum*, or Theory of the motions of Planets and Comets, &c, &c.

1746. The comet of this year was first observed at Lausanne, August 13, by M. des Chezeaux; and was seen afterwards by many other astronomers. It was then proceeding to its perihelion, to which it arrived not before February 8, 1747.

1748. M. Struick informs us, (Philosophical Transactions) that in the month of May, this year, three comets were visible, both at Amsterdam, and in other parts of Europe, on the very same night; of which there is no other instance in history: one of them was observed by F. Hallerstein, at Pekin in China, from April 26, to June 18, who says (Philosophical Taransactions abridged, vol. x.) ' The comet seen by us this year was very dismal, ‘for beſides its ſhining with a very obſcure and malignant light, it went in ſo deſert a path, and in ſuch an unfavourable ſky, that it could be obſerved but very ſeldom, and be compared with but a few ſmall ſtars not well known.’

1757. It is remarked by M. Montucla (Hiſtory of Mathematics) that near ten years had elapſed ſince any comet had appeared. A very uncommon circumſtance, if we may judge by the frequency of theſe phenomena for ſome centuries paſt, ſince comets have been ſo narrowly watched. The comet of this year, however, came kindly to relieve our impatience. It was accurately obſerved by Dr. Bradley, from September 12, to October 11, and again on the 18th and 19th of the ſame month. “When I firſt diſcovered this comet (ſays the doctor, Philoſophical Tranſactions, vol. 1.) it appeared to the naked eye like a dull ſtar of the fifth or ſix magnitude; but viewing it through a ſeven foot teleſcope, I could perceive a ſmall nucleus, (ſurrounded, as uſual, with a nebulous atmoſphere) and a ſhort tail, extending in a direction oppoſite the Sun.”

“It kept nearly at the ſame diſtance from the earth for about ten or twelve days together, after I firſt ſaw it; but its brightneſs gradually increaſed then, becauſe it was going nearer to the Sun. Afterwards, when its diſtance from the earth increaſed, though it continued to approach the Sun, yet its luſtre never much exceeded that of ſtars of the ſecond magnitude; and the tail was ſcarce to be ſeen by the naked eye."

The elements the doctor has given of this comet (adapted to Dr. Halley's general table for the motion of comets in parabolic orbits) will be ſufficient to enable future aſtronomers to diſtinguiſh it upon another return; but as they do not correſpond with the elements of the orbit of any other comet hitherto taken notice of, we cannot at preſent determine its period.

This comet was likewiſe obſerved by Mr. Klenkenberg at the Hague, who in a letter to Dr. Bradley ſays, "It appears very evident not only from my calculation, but from every other circumſtance of this comet, that it is not the ſame with that of the year 1682, which on certain accounts is very deſirable to be known; for both here, and in other parts of the Netherlands, there have been ſome people who have publiſhed mere conjectures; and have ventured (very minutely and exactly, as they pretended) about the time that this comet firſt appeared, to predict the return of that of 1682: but by the above obſervations the weakneſs of their pretenſions is evident; whereas, if this had proved to be the expected comet, they would have aſſumed to themſelves much undue praiſe, and have pretended to knowlege even ſuperior to the every-where much celebrated Newton, and Halley."

1758. Aſtronomers were much diſappointed this year in not finding Dr. Halley's prediction fulfilled: who in his Synopſis of comets has theſe words,—" and indeed there are many things which make me believe that the comet which Appian obſerved in the year 1531, was the ſame with that which Kepler and Longomontanus took notice of and deſcribed in the year 1607, and which I myſelf have ſeen return and obſerved in 1682. All the elements agree, and nothing ſeems to contradict this my opinion, beſides the inequality of the periodic revolutions: which inequality is not ſo great neither, as that it may not be owing to phyſical cauſes. For the motion of Saturn is ſo diſturbed by the reſt of the planets, eſpecially by Jupiter, that the periodic time of that planet is uncertain for ſome whole days together. How much more therefore will a comet be ſubject to ſuch like errors, which riſes almoſt four times higher than Saturn, and whoſe velocity, though increaſed but very little, would be ſufficient to change its orbit from an elliptical to a parabolic one. However, I am further confirmed in my opinion of the comets which appeared at the above periods, being the ſame; and let me alſo add, that in the year

1456, in the ſummer time, a comet was ſeen paſſing retrograde between the Earth and Sun, much after the ſame manner: which, though nobody made obſervations upon it, yet from its period, and the manner of its tranſit, I cannot think different from thoſe I have juſt mentioned. Hence I dare venture to foretel that it will return again about the year 1758."

As this is a point, and a period, of equally great importance in aſtronomy, we muſt not paſs them over too haſtily. Mr. Barker in a letter to Dr. Bradley, 1755 (inſerted in the Philoſophical Tranſactions of that year) has given, in twelve ſhort tables, the apparent path of this comet, ſuppoſing its perihelion any month in the year, with its accurate diſtance from the earth. But as no allowance was made for the diſturbance this comet might have met with, either from the planets, or other comets, in its path, it did not return within the period for which his tables were conſtructed.

But while all the world was big with expectation, and aſtronomers had turned night into day, in hopes of the accompliſhment of that prediction which was to confirm their favourite theory: while ſcoffers began to triumph in the hope that theſe ſtar-gazers were no greater conjurors than themſelves; and while the friends to ſcience began to tremble for the event: the profound and indefatigable M. Clairaut, one of the famous academicians who accompanied M. de Maupertuis in his voyage to the polar circle, remembering D. Halley had ſuggeſted that it was poſſible for the comet of 1682 to be impeded or accelerated in its courſe, by its approximation to Jupiter, went to work in order to diſcover by calculation, its approaches, not only to Jupirer, but to the reſt of the planets, and to find out their attractive powers over it. What a ſtupendous undertaking! but let him ſpeak for himſelf. 'The return of the comet of '1682, in the time preſcribed by the Newtonian 'theory, (advertiſement to his theory of comets, 'Par. 1760) is one of thoſe events which diffuſe 'new light upon the laws of nature, and which 'conſtitute a memorable aera in ſcience. An e'vent which has effectually diſſipated the laſt re'maining cloud which could poſſibly obſcure the 'ſyſtem of attraction.'

" In the year 1757, I had a mind to make a new application of the ſolution I had given of a famous problem ten years before *, to demonſtrate the univerſality of gravitation. The ſubject which afforded me this new application, was the comet

* This was the problem of three bodies, firſt applied to the theory of the Moon. Vide Principia, lib. iii. prop. 25. prob. 6.

of 1682, which was then expected to return according to Halley's prediction from Newton's theory. As the action of the great planets upon this comet might produce one or many years variation in its period, it rendered its return so uncertain that it was equally expected in 1757 or 1759 *;

* Though this has been asserted by Mr. Barker and many others, from the first edition of the Synopsis of comets published in 1705; yet it is not exact. Dr. Halley then indeed foretels the return of this comet in 1758. But many years afterwards, when he had carefully searched into the catalogues of ancient comets, and discovered that three others in the same order, and at like intervals of time, had preceded the three upon which his conjecture was founded, he began to be much more confirmed in his former opinion, of all these being one and the same comet. But, after accounting for some small diversity in their inclinations and periods from the action of Jupiter, which, by its attraction, alters the proper velocity of the comet when in its neighbourhood, the doctor adds, 'it is probable that its return will not be till after the period 'of seventy-six years or more, about the end of the year 1758, 'or the beginning of the next. *Circa finem anni* 1758, *vel initium proximi futurum.*' This puts the year 1757 quite out of the question.

N. B. The above is taken from Dr. Halley's Tabulae Astronomicæ, published in 1749, seven years after the author's death, and ten years before the accomplishment of his prediction, which he finishes by these remarkable words. 'You 'see therefore an agreement in all the elements of the three 'last appearances, (in 1531, 1607, and 1682) which would be 'next to a miracle if they were three different comets, or if it 'was not the approach of the same comet towards the Sun 'and earth, in three different revolutions round them. Wherefore, if according to what we have already said, it should 'return again abont the year 1758, candid posterity will not 'refuse to acknowlege that this was first discovered by an Englishman.'

I proposed therefore to find the true time when the expected comet would reach its perihelion."

"The labour upon which I entered was immense, and I was unable to arrive at any certain conclusion before the Autumn of 1758. I then thought it behoved me not to lose a moment, ere I acquainted the public and astronomers, with the result of my operations."

November 14, 1758, he presented to the Royal Academy of Sciences a memorial upon the subject and success of his enquiries. He there undertakes to prove, that the retardation of the expected comet, so far from injuring, would confirm the system of attraction, as it was a necessary consequence of the extent of that power. 'This is a question which has not hitherto been examined by 'geometricians: if it had, the result must always 'have been given conditionally. A body which 'passes into such remote regions, and remains out 'of sight during such long intervals, may be affected by causes wholly unknown to us; such 'as the action of other comets, or even by planets, 'too distant from the Sun ever to be perceived 'by us."

After this author had calculated all the disturbance that Jupiter might have occasioned to the comet during its three entire revolutions, a new difficulty occurred: he found it necessary to go through the ſame operations with regard to Saturn; the maſs of which planet being one third of that of Jupiter, might, *cæteris paribus*, produce one third of its effect: and that was ſufficient to merit a particular examination.

As to the other heavenly bodies in our ſyſtem, their maſſes not amounting to the hundredth part of thoſe of the two ſuperior planets, their effect is almoſt inſenſible.—He found that the action of Jupiter upon the comet, during the whole revolution of 1531 to 1607, had occaſioned a diminution of nineteen days in its period, which would not have happened by the mere force of the ſun; and at the ſame time had altered its elements ſo as to produce an acceleration of near thirty one days in the following period.

"Proceding afterwards to the Revolution from 1607 to 1682: The action of Jupiter turns out much more conſiderable; for it occaſions an acceleration of about 420 days, which added to the 31 reſulting from the action of the ſame planet during the preceding period, amounts in all to 451 days of diminution in the time of its period; which would not have happened merely by its inclination to the ſun."

"Now if we take the difference of thoſe two accelerations, in order to know how much ſhorter the ſecond period was than the firſt, it appears to be 432 days; which differs only 37 days from the time resulting from the observations."

"And this period appears to be still diminished by the action of Saturn. Indeed this diminution is not much, because the effects of Saturn's force are almost reciprocally destroyed in the two first periods."

"Hence we see that the theory gives within a month, the difference so remarkable between the two known revolutions of this comet. Now if we consider the length of these periods, the complication of the two causes of their irregularity, and the nature of the problem by which they are measured; this new demonstration of the Newtonian system will perhaps be found as striking as any one that has hitherto been given."

"By comparing, in like manner, the force of the action of Jupiter, during the second period of the comet, with that which will be terminated at its approaching return; I find the revolution about which we are at present interested will be 518 days longer than the preceding, occasioned by the action of Jupiter upon the comet, from its last mean distance to its perihelion: that is, for the last seven or eight years; an interval, during which there can hardly be more than fifteen days alteration."

"As to Saturn, the result of its action on the

comet is much more confiderable, compared with the two firft revolutions; for I find the prefent period protracted more than 100 days by it, independent likewife of its action fince 1751, and another fmall object which I have not had time to determine. From thefe confiderations, then, it appears to me that the expected comet ought to arrive at its perihelion, about the middle of the month of April next enfuing."

This is a long quotation, but the fubject of the memoir is curious, and the fuccefs of M. Clairaut, in determining fo nearly a point of fuch importance to aftronomy, and fo interefting to all lovers of fcience, makes us as defirous to render it public, as to augment that fame to which he has fo juft a title.

But M. Meffier, in an admirable memoir prefented to our Royal Society in 1765, (of which an excellent tranflation by D. Matty is publifhed in the tranfactions of that year,) has done juftice to Mr. Clairaut. And as this memoir, confifting of thirty pages, contains a minute and fatisfactory detail of the manner, in which the famous comet in queftion was firft difcovered at Paris, by M. Meffier, and afterwards obferved by him and M. de L'ifle; we fhall make no apology to our readers for giving them a long extract from a performance ſo fraught with entertainment and inſtruction.

1759. "In the predictions of the heavenly phenomena, which depend on the motion of the ſtars, two things are to be conſidered; viz. the time and place. As to the time, when the velocity, and direction of the ſtars in their motions, both apparent and real, are known; the time of their different appulſes and aſpects may always be foretold; and the accuracy of the calculations depends on the exactneſs with which their velocity and their ſeveral inequalities are aſcertained. Now it is well known, that all the former uncertainty, as to the exact time of the return of the comet foretold by Dr. Halley, was owing to the variations it muſt have undergone from its ſeveral ſituations, and approximations, to the planets, in its progreſs thro' the ſolar ſyſtem."

"Dr. Halley, who was firſt aware of the unequal returns of this comet in its former appearances, which he found to have been alternately of 75 and 76 years, was likewiſe the firſt who aſſigned their true cauſe. He aſcribed it, as I ſaid above, to the nearer, or more diſtant approaches of the planets of our ſyſtem: and having obſerved, that the comet we are ſpeaking of, came very near Jupiter in the ſummer of 1681, above a year before its laſt appearance, and remained ſeveral months in the neighbourhood of that planet, he judged that circumſtance alone ſufficient to have conſiderably retarded its motion, and prolonged the duration, of its revolution. Hence he concluded, that its return was not to be expected till the latter end of 1758, or the beginning of the next year."

" Dr. Halley obſerves, in confirmation of this opinion, that the action of Jupiter upon Saturn, is alone ſufficient to alter the duration of Saturn's period one whole month; and he adds, how much greater irregularities muſt not a comet be liable to, which, at its remoteſt diſtance, gets near four times farther from the Sun than Saturn; and whoſe velocity, in drawing near the Sun, needs but a very ſmall increaſe to change its elliptic into a parabolic curve."

" Dr. Halley does not determine more exactly the time of the return of the comet of 1682; neither could he do it, but by determining exactly the effect of the neighbourhood of Jupiter; which muſt very ſenſibly affect the velocity with which the comet was moving towards the ſun. Beſides, regard muſt be had, not only to this approach to Jupiter in 1681, but likewiſe to the other approaches to this, and all the other planets, which act more or leſs upon the comet, as they do upon each other. In ſhort, it was neceſſary to conſider all the different ſituations and diſtances of all the planets, with regard to the comet, during the whole of its laſt revolution; and even during the former ones, when the returns had been found to be unequal."

"What immenſe labour! and what geometrical knowlege did this taſk not require? M. Clairaut, of the Royal Academy of Sciences, undertook it; and his reſults differed but one month from the obſervation. No ſmall degree of exactneſs this, conſidering the immenſity of the object. In November 1758, he publiſhed his concluſion, which allowed about 618 days more for the period that was to end in 1759, than for the former; whence he inferred, that the comet muſt be in its perihelion, towards the middle of April. He added however, (Journal des Scavans, Jan. 1759) 'Any one may think with what caution I venture upon this publication, ſince ſo many ſmall quantities, unavoidably neglected by the methods of approximation, may very poſſibly make a month's difference, as in the calculation of former periods.' It accordingly proved ſo, the comet having reached its perihelion on the 13th of March in the morning. M. Clairaut has ſince publiſhed the methods and calculations, by which he has arrived at this concluſion."

"The impatience of aſtronomers, and their deſire to prepare for verifying this prediction of Dr. Halley, had put them upon enquiring for ſeveral years, in what part of the heavens this comet was likely to appear; but being ignorant of the exact time of its return, they could not determine the ſpot where it might be expected to be ſeen, but by making various ſuppoſitions as to the time of its perihelion. This Mr. Dirck of Klinkenberg, a famous aſtronomer in Holland, had attempted ſeven or eight years before; having taken the pains to calculate the principal points of fourteen different tracts, which the ſaid comet was to take, upon as many different ſuppoſitions relating to its paſſage thro' its perihelium, almoſt from month to month, from the 19th of June 1757, to the 15th of May 1758. Meſſrs. Pingré, and de la Lande, proceeded much in the ſame manner in the calculations they publiſhed in the memoirs of Trevoux, for April 1759, firſt and ſecond parts; with this difference, that the latter in their ſuppoſitions had taken narrower limits, and nearer to M. Clairaut's determination, who, as I ſaid before, had fixed the return of this comet to the middle of April."

"M. de Liſle, being curious of ſeeing the comet on its firſt return, as ſoon as it could be diſcovered by means of refracting, or reflecting teleſcopes, before it was viſible to the naked eye, thought he muſt proceed in a different manner from what other aſtronomers had done, to find out in what part of the heavens it muſt be looked for. He conſidered, that it was not neceſſary to know its place throughout its whole courſe, but only at the firſt moment of its appearance; becauſe, having once found it out, it would be an eaſy matter afterwards to trace it thro' its whole progreſs by obſervation and calculation."

"A full deſcription of this method is to be found in an ample memoir concerning this comet, which I have laid before the Royal Academy of Sciences, at Paris; and which no doubt will be printed in their collection, together with a northern hemiſphere, by means of which I have been enabled to look for this comet, in the very place of the ſky, where it ought to appear: and it was by the help of this planiſphere, that I actually diſcovered the comet from the Marine Obſervatory at Paris, on the 21ſt of January in the evening, after ſearching for it two years ſucceſſively, whenever the ſky would permit. The weather was extremely clear the 21ſt of January the whole day and evening. I ſeized this opportunity, and as ſoon as the ſtars were viſible after ſun-ſet, I examined through a Newtonian teleſcope of four feet and an half, thoſe places of the ſky, where the planiſphere ſhewed that the comet was to be expected."

"After much pains, I perceived about ſeven o'clock, a light reſembling that of the comet I had obſerved the year before in Auguſt, September, October, and the beginning of November *. I immediately made a configuration of this new light, with reſpect to the neighbouring ſtars, in order to examine the next night, whether it had any motion among the fixed ſtars. This light appeared pretty large, and in the middle I obſerved a nucleus, or bright ſpot, which was no proof as yet that it was a comet, as there are ſome nebulous ſtars with a bright ſpot in the middle."

" January 22, at the ſame hour as the day before, the ſky being equally clear, I again ſaw the ſame light with a four feet and an half teleſcope, and found it had ſenſibly changed its place; but its appearances were the ſame. From this ſecond obſervation I no longer doubted of its being a comet."

According to M. Meſſier, this comet had three ſeveral appearances above the horizon, which M. de Liſle, and he calculated, as ſoon as they had made their firſt obſervations, that is, as early as the month of February.

" The firſt appearance of this comet was in the evening, from January the 21 to February 14, when I ceaſed ſeeing it, by reaſon of its entrance into the rays of the Sun. The ſecond appear-

* See Mem. de l'Acad. Roy. des Scienc. Anno 1759.

ance was at the comet's getting clear of the rays of the ſun, in the morning, after the conjunction with that luminary, which was to take place a few days before its paſſage through the perihelion. I obſerved it in the morning from the 1ſt of April to the 17th, when it entered the rays of the ſun a ſecond time." During this ſecond appearance, the comet was much larger, and brighter than in the middle of February; and indeed it was but 18 days paſt its perihelion. Now it is well known that comets are much brighter after their perihelion, than at the ſame diſtance before it. "Beſides (ſays M. Meſſier) the comet after paſſing the perihelion, was as near again to the earth as on the 14th of February, when I loſt ſight of it at night. When I ſaw this comet again on the 1ſt of April, I could very plainly diſcern its tail, but could not aſcertain its length, becauſe of the morning twilight, which was then beginning, and ſoon encreaſed much: it filled the field of the teleſcope; and muſt have extended far beyond: according to what I have obſerved, the tail of the comet muſt have ſpread to more than 25 degrees; the nucleus was conſiderable, but not well terminated, and it apparently exceeded the ſize of ſtars of the firſt magnitude: it was of a pale whitiſh colour, not unlike that of Venus. The nebuloſity which ſurrounded the nucleus, and went on leſſening, ſhewed reddiſh colours; and theſe colours grew more vivid, towards the brighteſt part of the tail. The morning twilight, which increaſed apace, ſoon put an end to theſe appearances, and afterwards made the comet itſelf diſappear; however, I had been able to perceive it with the naked eye, when it was ſomewhat diſengaged from the vapours of the horizon."

"The third appearance of the comet was on the 29th of April in the evening, and I went on obſerving it till the 3d of June at night, when I ſaw it no more." During this laſt apparition, May 1, it appeared to the naked eye larger than ſtars of the firſt magnitude, the nucleus ſurrounded with a great coma. Its light was but faint, like that of the planets ſeen through the thick vapours of the horizon. It would have appeared brighter but for the light of the moon. In this laſt appearance of the comet above the horizon, it was in the ſextant, and was obſerved by moſt of the aſtronomers in Europe. The whole duration of its appearance was 134 days, reckoning from the 21ſt of January to June 3.

M. de la Lande's account of the return of this comet, (prefixed to his edition of Dr. Halley's aſtronomical tables, publiſhed in 1759, juſt after the departure of the comet) is very full, and ſatisfactory. We ſhall therefore preſent our readers with ſuch paſſages of that work as ſeem moſt intereſting, firſt premiſing, that M. Clairaut confeſſes himſelf obliged to M. de la Lande for aſſiſting him in his great work of calculating the diſturbances incident to the comet from its vicinity to Jupiter, &c. And M. de la Lande again, on his part, ſeems willing to participate this glory with Madam Lepaut, a lady who has long and ſucceſsfully been employed in aſtronomical calculations, to whom he acknowleged himſelf indebted for help in the part he had undertaken.

" The whole univerſe, ſays this author, has been witneſs to the accompliſhment of Dr. Halley's famous prediction, by the return of the comet of 1682, which deſcended to its perihelion May 13, 1759, after a period of 27937 days, or 76 years and 6 months."

" A German pamphlet publiſhed at Leipſic laſt January, and many printed letters from Germany aſſure us, that it was ſeen by a peaſant in the neighbourhood of Dreſden, ſo ſoon as the 25th of December 1758. An aſtronomer of the ſame country alſo obſerved it ſoon after, of which he gave information to ſeveral of his friends *. And M.

* M. le Monnier (Mercure de France, Apr. 1759.) remarks, that not only this comet had been firſt ſeen in Saxony, but the great one of 1680 had been ſeen there likewiſe two months

Meſſier diſcovered it at M. de Liſle's, 21ſt of January (as related above.)"

"The public was much ſurpriſed at this comet having no tail viſible to the naked eye, though it always had one in its former appearances. But for this many reaſons may be aſſigned. In the month of April, indeed, though the comet was near the Sun, yet it was very far from the earth—from whence it could only be ſeen during the twilight. Now, it is well known, that not only the twilight, but even the light of the moon is ſufficient to efface the tail of a comet. Hence we ſhould ceaſe to wonder that no tail appeared in the month of April—let us ſee now what was its poſition in 1607 and in 1682, when the ſame comet is ſaid to have appeared with a remarkable tail. September 20th 1607, Longomontanus ſaw it with a very long and denſe tail, which was 28 days before its perihelion; now ſuppo[illegible] the earth's diſtance from the ſun to be as 10, the comet was

before it was obſerved either in France or England. Occaſioned, according to this eminent aſtronomer, by the land to the eaſtward of that electorate being ſandy and dry. And as the eaſt winds bring few clouds, they have there calm weather and a clear ſky. "It is to be wiſhed, ſays M. le Monnier, that thoſe who inhabit climates where the ſky is more ſerene than ours, may have watched this comet as narrowly as has hitherto been done in France; to ſuch it will be viſible in the ſextant, that is to ſay, a little below Leo, till the end of July."

then only 2 of thoſe parts diſtant from the earth, and 8$\frac{1}{2}$ from the ſun. Auguſt 29, 1682, M. Picard ſaw the comet with a tail 30 degrees long; Hevelius allowed it only 16 degrees: but this was 16 days before its perihelion; it was then diſtant from us 3$\frac{1}{2}$ 10ths and from the ſun 6$\frac{1}{2}$."

" Hence in both caſes, there is a more favourable combination in its diſtance, both from the ſun and the earth, than when it laſt appeared, which is ſufficient to explain the different figure it made. We ſhould then treat with all due contempt every ſuſpicion of this comet not being the ſame as that of 1682: its inclination, perihelion, nodes, diſtance from the ſun, motion *, and even its late arrival occaſioned by the attractions of Jupiter and Saturn, which ſo well agree with calculation; all theſe circumſtances amount to ſo full and ſtriking a demonſtration, that I am aſhamed to ſtop a moment at ſuch difficulties. However, as the academy always publiſhes the reſult of its labours; and as doubts, however groundleſs, always occaſion a ſuſpenſion in the progreſs of the human mind, I thought I ſhould be excuſed by men of ſcience if I tried to remove objections which, per-

* All theſe may be ſeen and compared by any one who will take the trouble to inform himſelf of the meaning of theſe terms, and to caſt his eye over the following table, conſtructed by the Abbe de la Caille. Lecons d'Aſtronomie 1761.

TABLE

OF THE

ELEMENTS of the ſeveral REVOLUTIONS of HALLEY'S Comet.

Year of Appearance.	Place of the Aſcending Node.				Inclination of the Orbit.			Place of Perihelion.				Log. of the Perihelion Diſtance.	Paſſage thro' the Perihelion, mean time at Paris.				Motion.	Orbit by whom Calculated
	S.	D.	M.	S.	D.	M.	S.	S.	D.	M.	S.			D.	H.	M.		
1456	1	18	30	0	17	56	0	10	1	0	0	9. 767540	June	8	22	10	Retrograde	Pringre.
1531	1	19	25	0	17	56	0	10	1	39	0	9. 753583	Auguſt	24	21	27	Retr.	Halley.
1607	1	20	21	0	17	2	0	10	1	16	0	9. 768490	October	26	3	59	Retr.	Halley.
1682	1	20	48	0	17	42	0	10	1	36	0	9. 765296	Septem.	14	21	31	Retr.	Halley.
1759	1	23	49	0	17	39	0	10	1	16	0	9. 766039	March	12	13	41	Retr.	La Caille.

haps, with ſome may gain credit, however ill founded."

The moſt important objection, as to the return of this comet, ariſes from the inequality of its periods, which were as follows: that from Auguſt 25, 1531, to the 26th of October 1607, was performed in 76 years, and two months; that from October 26, 1607, to September 14, 1682, was rather leſs than 75 years; and its laſt period from the 14th of September 1682, to the 3th of March 1759, which was the longeſt of all, was 76 years and ſix months; or 27,937 days, amounting to 583 days more than in the preceding period.

Dr. Halley was aware of theſe differences, and at firſt confeſſed himſelf to be a little ſtaggered by them, nor would he have had the courage to pronounce its return ſo poſitively, if hiſtory had not informed him, that comets had appeared in 1456, 1380 and 1305, which put their identity out of all doubt.

Theſe appearances happening alternately in ſeventy-five and ſeventy-ſix years, and as the preceding period was only of ſeventy-five years, it was natural to ſuppoſe that the next would amount to ſeventy ſix. But as the difficulties ariſing from theſe inequalities in the periods have been foreſeen and obviated by Dr. Halley, we cannot do better than to insert his own words.

" Perhaps some may object to the diversity of their inclinations and periods, which is greater than what is observed in the revolutions of the same planet; seeing one period exceeded the other by more than the space of one year, and the inclination of the comet of the year 1682, exceeded that of the year 1607, by twenty-two entire minutes. But let it be considered what I mentioned at the end of the tables of Saturn, where it was proved that one period of that planet is sometimes longer than another by thirteen days; and that is evidently occasioned by the force of gravity tending towards the centre of Jupiter, which force indeed in equal distances is only the thousandth part of that force tending to the Sun itself, by which the planets are retained in their orbits. But by a more accurate computation, the force of Jupiter towards Saturn, for example, in the great conjunction as they call it, January 26, in the year 1683, was found to be to the force of the Sun upon the same Saturn, as 1 to 186; the sum of the forces therefore is to the force of the Sun, as 187 to 186. But at the same distance from the center, the periodic times of bodies revolving in a circle are in the subduplicate ratio of the forces with which they are urged: wherefore the gravity being increaſed by 186th part of itſelf, the periodic time will be ſhortened by about the 374th part, that is, by a whole month in Saturn. How much more is a comet liable to theſe errors, which makes its excurſion near four times higher than Saturn: and whoſe velocity being increaſed by leſs than the 120th part of itſelf, would change its elliptic orbit into a parabolic trajectory."

"But it happened in the ſummer of 1681, that the comet ſeen in the following year, in its deſcent towards the Sun, was in conjunction with Jupiter in ſuch a manner, and for ſeveral months ſo near him, that during all that time it muſt have been urged likewiſe towards the centre of Jupiter with near the 50th part of that force by which it tended towards the Sun: whence, according to the theory of gravity, the arc of the elliptic orbit, which it would have deſcribed had Jupiter been abſent, muſt be bent inwards towards Jupiter in an hyperbolic form winding, and have aſſumed a kind of curve very compounded, and as hitherto not to be managed by the geometers; in which the velocity and direction of the moving body, in proportion to the cauſe, would be very different from what it otherwiſe had been in the ellipſes."

"Hence a reaſon may be aſſigned for the change of its inclination: for as the comet in this part of its path had Jupiter on the north almoſt in a perpendicular direction to its path, that portion of its orbit muſt be bent towards that quarter; and therefore its tangent being inclined to a greater angle towards the plane of the ecliptic, the angle of the inclination of the plane itſelf muſt be neceſſarily increaſed. Beſides the comet continuing long in the neighbourhood of Jupiter, after it had come towards him from parts much more remote from the Sun with a ſlower motion, and now being urged with the joint central forces of both, muſt have acquired more accelerated velocity, than it could loſe in its receſs from Jupiter, by forces acting a contrary way, its motion being more ſwift, and the time being leſs." (*Tabulæ Aſtronomicæ*)

When the comet of 1682 deſcended towards the Sun and became viſible, Europe had ſcarce recovered from the terrible panic into which it had been thrown but eighteen months before by the great comet. However, this was comparatively too inconſiderable to be much regarded, for it was little imagined then, that the leaſt of the two would become the moſt intereſting, and that it would be for ever celebrated by poſterity for having taught mankind how to know all the reſt. But however inferior to the other this comet may have appeared in vulgar eyes, aſtronomers obſerved it with the greateſt attention. Hevelius at Dantzick, Kirch at Leipſic, Flamſtead and Halley in England, Zimmerman at Nurenburg, Baert at Toulen, Montanori at Padua, and Picard, Caſſini and la Hire at Paris. This liſt of names will ſuffice to ſhew that there can be no ſcarcity of good obſervations upon this comet during that appearance.

In 1607 it was obſerved by the famous Kepler, who publiſhed his obſervations together with his general theory (*de Cometis Libelli* 3, *autore Joanne Keplero, auguſtæ vindelicorum* 1619.) The 16th of September old ſtile, the ſky being very clear, Kepler firſt ſaw this comet upon the bridge at Prague, and though it had no tail when he firſt diſcovered it, yet afterwards it had one of a conſiderable length and ſplendor. It was likewiſe obſerved by Longomontanus, September 18, (*Aſtron. Danicæ appendix, Amſt.* 1640.) he ſays it appeared as large as Jupiter, though with a very obſcure and pale light; that the tail was pretty long and more denſe than the tails of comets uſually are, but as pale in colour as the comet itſelf.

In the preceding revolution of 1531, we find our comet obſerved by the aſtronomer Appian at Ingoldſtadt, the ſame who firſt remarked that the tails of comets were always in an oppoſite direction to the Sun: which to him was an evident proof that the Sun was the cauſe of ſuch eruptions.

In 1456, there was a very remarkable exhibition of the ſame comet. *Cometa in auditæ magnitudinis toto menſe Junii cum prælonga Cauda, ita ut duo fere ſigna cœli comprehenderit.* (*Theatrum Comet:*)

It is difficult to comprehend how the comet whoſe tail was ſo inconſiderable in its laſt appearance, ſhould in this have one of ſixty degrees: but M. de la Lande in his theory of comets, p. 127. accounts for this difference in the following manner. " I find, ſays this active aſtronomer, that if the comet reached its perihelion in the beginning of June, it ought to have appeared at night towards the middle of the month with ſixty degrees of elongation and a very northern latitude, its diſtance from the earth being leſs than the ſemidiameter of the Sun: ſo that in this poſition, which of all others is the moſt favourable, it muſt have appeared in all the ſplendor allowed to it by the old chronicles. Perhaps by *duo ſigna*, they only mean the extent of two conſtellations, which is often much leſs than two ſigns of the ecliptic."

In 1379 and 1380 we find two comets mentioned by Alſtedius and Lubienietzki, but without any particulars as to the time or form of their appearance.

In 1305, our comet again appears, according to the hiſtorians of that time, in all its terrors.

Cometa horrendæ magnitudinis visus est circa ferias paschatis, quem secuta est pestilentia maxima; it is very likely that the horror occasioned by the plague had augmented the terrible impression left by the comet; however, upon calculation, it does appear that the comet must this year have passed very near the earth.

The history of this comet might be traced much higher by consulting Eckstormius, Riccioli, Alstedius, and Lubienietzki. Among the four hundred and fifteen comets mentioned by this last writer, we find one for the year 1230, which appears to be the very comet in question; another in 1005, three periods before; it is found in 930, and higher up in the year 550, marked by the taking of Rome by Totila. All the historians of the empire speak of a great comet in the year 399, which may have been the same. *Cometa fuit prodigiosæ magnitudinis, horribilis aspectu, comam ad terram usque dimittere visus.*

In 323, that is to say, seventy-six years before, a comet also appeared in Virgo; and in short it would be easy to mount, without quitting the same periods, as high as 130 years before Christ, when, according to Justin, one appeared at the birth of Mithridates. But, in these early periods, there would be great danger of meeting with some of those fabulous comets with which it was thought necessary perhaps to embellish every famous reign: and it must be confessed too, that equal intervals between the different apparitions of comets, are not alone sufficient to prove their identity: such equalities may indeed contribute towards the support of a demonstration founded on an agreement in their motions, and a perfect correspondence in the other circumstances of their appearance, but greater stress must not be laid upon them: for these compilations were not formed with the same care and exactness, which would have been bestowed upon them, if, when they were made, it had been suspected what advantages were to be derived from them. Lubienietzki seems to have had no other view than to compare the events subsequent to the appearance of comets, in order to prove that they have presaged *nothing*: just as his predecessors, among whom was the good father Riccioli, had compiled them in order to prove them to be *inauspicious augurs*,

Riccioli, in his Almagest. published in 1651, enumerates 154 comets to be found upon record in history, the last of which appeared in 1618. But in the great work of Lubienietzki*, (a Polish gentleman descended from the Sobieski family, but who being tainted by Socinianism was forced

* Theatrum Cometicum, Amst. 1668.

to quit his country,) where not a ſingle hiſtorical circumſtance relative to comets is omitted, the number in 1665 amounted to 415. "Since that time, ſays M. de la Lande, in 1764 (Aſtronomie, 2 vols. 4to) they are increaſed to 450. But of all theſe appearances, no comet had its path aſtronomically deſcribed till 1264 *, and the number of thoſe which have been obſerved, with ſufficient accuracy to determine their orbits, is reduced to fifty one, excluſive of the comet of 1531, 1607, 1682 and 1759 which is allowed to be only different returns of one and the ſame comet."

It ſhould be remembered that though every meteor and ſtrange appearance in the heavens was by the ancients called a comet, and that many of thoſe which were intitled to that appellation, were the ſame comets ſeen at different revolutions; yet it may eaſily be ſuppoſed, that in every age, and eſpecially in the early ones, many comets have appeared concerning which hiſtorians have been ſilent, as well as many others, which on account of their diſtance, or of cloudy ſkies, have not been viſible to the inhabitants of the globe.

We muſt not wonder then, if among the 415 comets mentioned by Lubienietzki, there are near

* Tractatus Fratris Egidii de cometis.

400 from which nothing poſitive can be conclud- ed. But whatever uncertainty there may be in theſe remote periods, we have four returns of one comet perfectly well aſcertained, which joined to that of 1759, put the theory of this comet out of the reach of cavil, and conſtitute the greateſt triumph of aſtronomy and the higheſt glory of the human mind.

Dr. Bevis obſerved this comet in London, May 1 and 2, and exultingly ſays (Phil. Tranſ. vol. 51.) " I think I may now venture to pronounce this to be the ſame as the comet of 1682; and am about making out its future track. If I pre- ſume rightly, it will in a ſhort time become in a manner ſtationary, but diminiſh very faſt both in ſize and light, the earth and it, receding from each other almoſt in a right line. It is at this time about four times nearer the earth than the Sun is."

Mr. Muncley likewiſe obſerved it at Hampſtead, April 30, May 1, 2, 5, and 6. ' It is a lu- ' minous appearance, ſays he, very evident to ' the naked eye (notwithſtanding the light of the ' Moon, within two or three days of her qua- ' drature) yet rather dim than ſplendid, large, but ' very ill defined, &c.'

We cannot quite this article without mention- ing, that though the period of this comet is the ſhorteſt of any yet diſcovered, (Dr. Halley calls it the Mercury of comets) its aphelion, or greateſt diſtance from the Sun, is thirty-five times greater than that of the Earth, and four times greater than that of Saturn, the moſt remote of any of the planets.

Indeed this comet, ſo big with conſequences, ſeems, it muſt be allowed, very diminitive as to ſize, compared with many others: however, no one point in aſtronomy ever engaged the attention of ſo many great aſtronomers as the return of this comet. Newton, Halley, Maupertuis, Clairaut, de Liſle, le Monnier, la Caille, Meſſier, la Lande, Pingré, &c. have been indefatigable in obſerving and calculating its courſe. There was a contro-verſy among the French aſtronomers concerning the methods of finding it, and the exact time of its perihelion; but they and all the aſtronomers in Europe were unanimous in pronouncing it to be the ſame comet which appeared in 1682; and here we cannot help repeating, for the honour of aſtronomy and of the Engliſh nation, that this co-met was firſt calculated, and its return predicted by the great Dr. Halley, in confirmation of the theory of the illuſtrious Sir Iſaac Newton.

1760. The comet of this year was diſcover-ed, in the conſtellation Orion, at Cambridge, by Dr. Maſon, and at Paris by M. Meſſier, on the ſame night, and at the ſame hour: namely, January 8, about nine o'clock in the evening. January 9, the late Mr. Short, Mr. Munckley of Lincoln's-Inn, and Mr. Day at Lowick, Northamptonſhire, ſeverally obſerved it. They all ſpeak of the extreme rapidity of its motion, and of the body being ill defined; but of the tail, thoſe who ſaw it in England, ſay nothing, except Mr. Short, who, after remarking that its motion was to the weſtward with a conſiderable velocity, ſeemingly about two degrees in a day, which was nearly at the rate the great comet moved, when it was firſt ſeen in the end of the year 1743, adds, 'This 'comet is very viſible to the naked eye, though 'I could perceive nothing of a tail, and therefore 'I conclude it is going down to the Sun.' But the account of this comet in the *Mercure de France*, of January 1760, ſays, the tail has an eaſtern direction, and is about four degrees long, but ſcarcely viſible to the naked eye. The late Dr. Stukely ſeems to have thought this comet was the ſame as that of 1664. But M. Barker (Phil. Tran. vol. 52.) rather diſcourages that opinion. " The comet of 1664, ſays he, might have appeared nearly in the ſame place this was ſeen, with a ſwift motion of pretty many degrees in a day, as a retrograde comet in oppoſition to the Sun generally has; but, I think, would not have been near enough to have moved a degree in an hour as this did; and I think it would have been alſo a larger, and continued longer than this; for in 1664, it was ſeen four months, and when far diſtant from the earth; and in the poſition it muſt have been in laſt January, would hardly have gone farther back than the beginning of Gemini, in ſmall north latitude, and is, I believe, one of the largeſt comets."

But M. Meſſier, who firſt diſcovered in France the three laſt mentioned comets, ſeems to have found this year a fourth comet, concerning which he communicated his obſervations to the Royal Academy of Sciences. This comet, wholly different from that of which we have been ſpeaking above, was firſt ſeen by M. Meſſier, January 26, at which time it could be perceived only through a foot reflecter, though ſoon after, with great difficulty, by the naked eye. Its nucleus appeared pretty clear, and well defined, through a 4 ½ foot Newtonian teleſcope. The day of its diſcovery, it was ſituated between the conſtellations *Crater* and *Hydra*. February 4, it was viſible in *Leo*: and on the eighth appeared to the naked eye equal to a ſtar of the third or fourth magnitude, and was very brilliant, having a tail (viſible indeed only through a teleſcope) of many degrees in length, with a weſtern direction.

1762. The comet which appeared this year, is ſuppoſed by Mr. Struick and M. Pingré, who both obſerved it, and compared their obſervations with thoſe of other aſtronomers, to be the ſame with that which appeared 169 years before, viz. in 1593. Its courſe was direct, and it arrived at its perihelion, May 28. In conſtructing the elements of this comet, a remarkable ſingularity occurred to M. Pingré: he found that it had paſſed eleven times nearer the Sun than the Earth does when it is in its perihelion; and likewiſe, that though it was ſeen a very few days after its perihelion, and might be expected to have equalled the celebrated comet of 1680 in ſplendor, yet it did not exceed in brightneſs a ſtar of the third magnitude, its tail at the ſame time not extending above four degrees. M. Pingré therefore ſuppoſes it to have been very ſmall, and that its atmoſphere was not qualified to abſorb or attract, according to M. Mairan's ingenious ſyſtem, a ſufficient quantity of thoſe luminous particles, which, ſays M. Mairan, compoſe the ſolar atmoſphere*.

1764. Two comets are this year diſcovered by the vigilant and perſpicacious M. Meſſier. Theſe ſtars by being ſo numerous, will ſoon ceaſe to be

* This article is extracted from the Hiſt. de l'Acad. des Sciences, for the year 1763, in which volume M. Bailly gives ſeveral obſervations on the ſame comet.

regarded with the ſame wonder as formerly; however, it muſt be owned, that the frequency of theſe diſcoveries is in a great meaſure owing to the uſe and improvement of teleſcopes; without which, we ſhould know no more of many that have lately been ſeen, than our ſhort-ſighted forefathers did of the ſatellites of Jupiter and Saturn. As to M. Meſſier, he is ſo conſtantly on the look-out, and ſo dexterous in diſcovering them, that it would incline one to believe with Caſſini, that there was really a zodiac of comets, and that M. Meſſier alone knew its place and limits in the heavens.

The firſt comet of this year was diſcovered at the Marine Obſervatory at Paris, March the 8th, in the conſtellation of *Piſces*, and obſerved till the 15th of the ſame month; its motion was retrograde, and it ſeemed, in ſize, equal to a ſtar of the fourth magnitude.

The ſecond comet was at firſt diſcovered with the naked eye, April the 8th, near the *Pleiades*, and promiſed to become conſiderable. The next day the tail was ſix or ſeven degrees long, and the nucleus equally luminous with ſtars of the third magnitude. It was however viſible only till the 12th. M. Meſſier ſent tables of the places of theſe comets to the Royal Society this year, together with a calculation of the elements of their orbits by M. Pingré.

The ſecond of thoſe two comets was diſcovered at Louiſburg, in the iſland of Cape Breton, April the 7th, which was one day ſooner than even M. Meſſier had ſeen it at Paris. In this obſervation, made by Captain Holland, the tail of the comet appeared perpendicular to the horizon, with its head towards the ſun.

Mr. Brice obſerved the ſame comet at Kirknewton, April the 10th. It was then deſcending towards the ſun, at the rate of about ſix degrees in the ſpace of twenty-four hours. To this gentleman's account, in the Philoſophical Tranſactions, is prefixed a plate of the appearance of the comet.

1767. The firſt intelligence we had of this comet in England, was from the indefatigable M. Meſſier, aſtronomer, keeper of the journals, plans and maps belonging to the marine of France, who diſcovered it the 8th of Auguſt, about 11 in the evening, in the conſtellation Aries, between the 24th, 25th and 31ſt ſtars of that conſtellation in the Britiſh catalogue. On the 14th and 15th of the ſame month it appeared very diſtinctly, having a tail about ſix degrees in length, This information was inſerted in the St. James's Chronicle, Auguſt 25, which ſet us all to work in order to find it here. Not many days had elapſed, ere it was ſeen by all who were poſſeſſed of teleſcopes, and in the beginning of September, it was viſible to the naked eye, about three o'clock in the morning, in the conſtellation of Taurus, with a tail fifteen degrees long. The body of this comet became more conſiderable to our view, till the middle of September, when the tail was of an enormous breadth, and extended to upwards of 40 degrees in length.

The public is, doubtleſs, much obliged to Mr. Dunn, for his obſervations, which appeared ſo frequently in the news papers, though he put them in a great fright for our beautiful morning and evening ſtar, the planet Venus, which, "he "thought likely to receive a *bruſh* from the co- "met's tail." However, he did not ſuffer their anxiety to continue long, but ventured two or three days after to pronounce Venus out of danger.—But may we not ſuppoſe that the whole ſolar ſyſtem, that is to ſay, our ſun—with its ſix planets, ten moons, and comets, as yet unnumbered, are ſo combined together, ſo dependent one on another, and ſo much one family, that neither Venus, the Earth, nor any one part of this ſyſtem can ſuffer alone, as ruin to one would, perhaps, be ruin to all.

From nature's chain whatever link you ſtrike,
Tenth or ten thouſandth, breaks the chain alike.

It has been ſaid above, that 450 comets are recorded in hiſtory. Now if we conſider, that moſt of theſe comets deſcend more or leſs, into the ſphere of the orb of the earth; and that out of thoſe whoſe orbits have been calculated, there are only ſix whoſe leaſt diſtance from the ſun, exceeds that of the earth—yet ſtill, no accident has happened that we can trace in the hiſtory of the moſt remote ages of the world:—May we not ſuppoſe, that room enough is aſſigned in infinite ſpace, by infinite power, for theſe orbs to move in, without falling foul of each other, as if left to the guidance of blind and blundering chance; and that we have nothing to apprehend from ſhort-ſighted predictions, or fanciful hypotheſes.

It is therefore to prevent too great diſturbances in the motions of comets from the action of the planets and other comets, ſays Sir Iſaac Newton, that while the planets revolve all of them nearly in the ſame plane, the comets are diſpoſed in very different ones, and diſtributed over the whole ſyſtem.

Maclaurin, too, has an admirable reflection upon this ſubject, in his paraphraſe on Sir Iſaac Newton's Principia. Speaking of the fatal effects that ſeemed poſſible to happen from the near approach of the great comet of 1680 to the earth, ſays, " it is not to be doubted but that, while ſo many comets paſs among the orbits of the planets, and carry ſuch immenſe tails along with them, we ſhould have been called by very extraordinary conſequences, to attend to theſe bodies long ago, if their motions in the univerſe had not been at firſt deſigned and produced by a being of ſufficient ſkill to foreſee their diſtant conſequences."

The preſent comet totally diſappeared about the 16th, being immerged in the rays of the ſun, paſſing with great rapidity to its perihelion, from whence we are now (October) impatiently expecting its return.

By an article from Paris we are likewiſe informed that it became inviſible there the ſame day on which it retired from us.

" The late comet, ſo much talked of, was diſcovered at Paris the 8th of laſt Auguſt, by that indefatigable aſtronomer, M. Meſſier, whoſe aſſiduity and dexterity in obſerving the heavenly bodies, have long ſince deſerved the higheſt praiſes from the learned; and he was honoured on this preſent occaſion with a letter, wrote to him upon the ſubject, by one of the greateſt monarchs in Europe, and geniuſes of this age. The accurate obſerver followed obſerving the courſe of the comet till the 16th of September, when it ceaſed to be viſible by its approaching to the ſun, being then near the alpha of Hydra. Its elements have been calculated by M. de la Lande, upon three equidiſtant obſervations of M. Meſſier, made on the 14th, 21ſt, and 28th of Auguſt, and are as follows:"

"Inclination of the orbit, 73 deg. 15 min. Deſcending node, 11 ſign. 26 deg. 23 min. Perihelion, 6 ſign. 11 deg. 28 min. Paſſage at the perihelion the 1ſt of October, at 9 and 22 min. Diſtance of the perihelion, 0,03104; that is to ſay, 32 times nearer to the ſun than our globe ever is. From this it appears, that this comet does not reſemble any of the 57 comets we know of; only the two ſeen in the years 1680 and 1689, did approach neareſt to the ſun. The laſt of them ſhould have ſome likeneſs to this, were it not for the great difference between the diſtances of their reſpective perihelions, which take away any ſuſpicion of this being the ſame."

According to the perihelion diſtance given to this comet by M. de la Lande, it might have been expected to return ſooner than has been found by experience, as the velocity it would have acquired by ſuch a near approach to the Sun, would have been accelerated, and its trajectory diminiſhed.

As this comet, in its way to the Sun, ſets weſtward of that luminary, it will riſe from its rays on the other ſide; namely, eaſtward of the Sun.

There is a popular diviſion of comets into three kinds; namely, *tailed*, *bearded*, and *hairy* comets, though this diviſion rather relates to the different circumſtances and ſituations of the ſame comet, than to the phenomena of ſeveral.

Thus, when the light is weſtward of the Sun, and ſets after it, the comet is ſaid to be *tailed*, becauſe the train follows it in manner of a tail.

When the comet is eaſtward of the Sun, and moves from it, the comet is ſaid to be *bearded*, becauſe the light marches before it in manner of a b ard.

And laſtly, when the comet and the Sun are diametrically oppoſite, (the Earth between them) the train is hid behind the body of the comet, except a little that appears round it, in form of a *border of hair*.*

The tails of comets are always on the ſide oppoſite to the Sun. This was firſt diſcovered by Appian, and has ſince been conſtantly confirmed by obſervation. They are beſt ſeen, and appear longeſt in ſouthern climates, where the air is pure and ſky ſerene. The comet of 1759 appeared at Paris almoſt without a tail, and in England

* From this laſt appearance the word comet is derived; as Κομητης, *Cometa*, comes from Κομη, *Coma*, a head of hair.

entirely without one. At the former place, it was with great difficulty that a ſlight trace of one could be diſtinguiſhed of only one or two degrees in length; whereas at Montpelier, M. de Ratte found it to have one, April 29, of 25 degrees in its whole length, and 10 degrees of it extremely luminous; but M. de la Nux ſaw that comet, at the iſle of Bourbon, with a tail much more conſiderable; for the ſame reaſon that the zodiacal light is always viſible there, and extends to above 100 degrees in length.

But there have been comets, whoſe diſk was as round, as well defined, and as clear as that of Jupiter, without either tail, beard, or coma; ſuch was one of the comets that appeared in 1665, and, according to Caſſini, that of 1682. Hence the tail of a comet muſt not always be regarded as its neceſſary appendage, or principal characteriſtick.

Moſt comets are viſibly ſurrounded with an enormous atmoſphere, often riſing ten times higher than the nucleus, or ſolid body of the comet. Sir Iſaac Newton ſuppoſes it to be owing to the atmoſphere of a comet, that the nucleus is uſually ſo ill defined; the moſt lucid parts of which not being above a ninth or tenth part of the whole breadth. In obſervations upon comets, it is common to meet with accounts of bright ſpots in the middle of the nucleus, when in fact it should seem that the bright spot only was the nucleus, and the rest the atmosphere of the comet.

M. de la Lande supposes very ingeniously, that as comets are destined to pass from the most dreadful rarefraction and heat imaginable, to a cold density beyond conception, they are provided with these immense atmospheres, not only to protect them from such destructive excesses, but likewise to support and foment circulation, fluidity, motion, and life.

It has been remarked that comets have different phases, like the Moon; and it was observed by Cassini, in the year 1744, that the body of that comet was horned, shewing only half its disk.

The present comet, seen through a good telescope, seems more to resemble a small Moon than a fixed star. The nucleus, or body of it, is large, but ill defined. The phaenomena of the tail seems so much to favour an ingenious conjecture in the Monthly Review, (Oct. 1767, p. 253.) that we cannot resist quoting it.

The book under examination is Dr. Priestley's History of Electricity: " Signior Beccaria has, with great ingenuity, mixed sometimes with a little spice of agreeable extravagance, the frequent concomitant of genius, ranged almost all the meteoric phenomena under the banners of electricity; from the *Will o'-th'-Wisp* up to the *Aurora Borealis.*

Had we room or inclination to theorise on this subject, at the same time that, with other electricians, we allowed the electric fluid to be the cause of this last phenomenon, we should be for extending its connections still further, and attempt to shew the possibility, at least, of its near relation to, if not its identity with, that luminous matter which forms the solar atmosphere, and produces the phaenomenon called the *Zodiacal Light;* which is thrown off principally, and to the greatest distance from the equatorial parts of the Sun, in consequence of his rotation on his axis, extending visibly, in the form of a luminous pyramid, as far as the orbit of the Earth; and which, according to M. de Mairan's ingenious, and, at least, plausible hypothesis, falling into the upper regions of our atmosphere, is collected chiefly towards the polar parts of the Earth, in consequence of the diurnal revolutions, where it forms the *Aurora Borealis.* It would, we think, be no very bad hypothesis which should unite these two opinions, by considering the Sun as the fountain of the electric fluid, and the zodiacal light, *the tails of comets*, the *Aurora Borealis*, lightening, and artificial electricity, as its various, and not very dissimilar modifications." Indeed the appearances of the tail of this comet resembled electrical coruscations, more than any thing of which we have an idea, but moſtly that produced *in vacuo*; as the flame ſeemed, through a teleſcope, perpetually to ſhoot out in ſtrait lines, of a pale ſilver hue, lengthening and ſhortening at each inſtant, and forming frequently ſome of the configurations which the *Aurora Borealis* aſſumes.

There has been lately publiſhed a work by Dr. Hamilton of Dublin, under the title of Philoſophical Eſſays; in which this idea ſeems extremely well developed. The ſubject and ſubſtance of the Doctor's ſecond eſſay is ſo full to our purpoſe, that we ſhall condenſe it into an epitome, and preſent it to our readers.

Dr. Hamilton's Eſſay has for title, Obſervations and conjectures on the nature of the *Aurora Borealis*, and *the tails of Comets*.

The author differs from Sir Iſaac Newton, concerning the nature of the tails of comets; and endeavours to prove that they are compoſed of a lucid, or ſelf-ſhining ſubſtance, and not a mere cloud or vapour, illuminated only by the Sun. This luminous matter he ſuppoſes to be the ſame with that which cauſes the Aurora Borealis, and the phaenomena of electricity.

“ The great body of luminous matter which appears in an *Aurora Borealis*, ſays this Author, being ſo very extenſive, and ſometimes ſo very bright, muſt be viſible to a ſpectator at a conſiderable diſtance from the Earth, and ſhaded from the Sun's light; and ſuch a ſpectator would then ſee the earth attended by a train of light in the form of a *tail.*"

" Electric matter appears to be of the ſame kind of ſubſtance which forms the *Aurora Borealis*, and *the tails of Comets;* by its having alſo that remarkable property of letting the rays of light paſs through it, without having any ſort of effect upon them." " Now the extraordinary rarity of comets tails may be collected, ſays Sir Iſaac Newton, from the ſtars ſhining through them; for the ſmalleſt ſtars are obſerved to ſhine without loſs of ſplendor through tails which are of an immenſe thickneſs." (*Principia*, p. 513, edit. 2.) Dr. Hamilton has given to comets a quite different employment from that allotted to them by Sir Iſaac Newton, who made them water-carriers, loading them with vapours and moiſture, to ſupply the loſſes of the ſeveral parts of the ſolar ſyſtem through which they were deſtined to paſs. But the Doctor, on the contrary, ſuppoſes it their buſineſs to collect and bring back to the Sun and planets, the electric fluid which is conſtantly flying off into the higher regions of the heavens, beyond the orbit of Saturn. " We ſee this fluid riſes from the earth into the atmoſphere, and is probably going off from thence, when it appears in the *Aurora Borealis*. And as this electric matter, from its vaſt ſubtilty and velocity, ſeems capable of making great excurſions from the planetary ſyſtem, the ſeveral comets in their long excurſions from the Sun, in all directions, may overtake this matter, and attracting it to themſelves, may come back replete with it, and being again heated and excited by the Sun, may diſcharge and diſperſe it among the planets, and ſo keep up a circulation of this matter, which we have reaſon to think neceſſary in our ſyſtem."

This does not ſeem far from Sir Iſaac Newton's own opinion: for after ſuppoſing that the aqueous particles thrown off from comets are taken up by the planets, as a ſupply of moiſture, he adds, 'I ' ſuſpect, moreover, that that ſpirit which is the ' leaſt, but the moſt ſubtile, and the beſt part of ' our air, and is neceſſary for ſupporting the life ' of all things, comes chiefly from the comets.'

Dr. Halley, ſo long ſince as the year 1716, in his deſcription of a remarkable *Aurora Borealis*, ſays, "That the great ſtreams of light ſo much reſembled the tails of comets, that at firſt ſight they might well be taken for ſuch;" and afterwards adds, "This light ſeems to have a great affinity to that which the effluvia of electric bodies emit in the dark." This was the doctor's conjecture;—but both he, and Sir Iſaac Newton, had ſo much of vaticination in all they ſaid, that their conjectures are found by poſterity to be little leſs than certainties.

But this ſtriking ſimilarity between the appearances of the tail of the comet in a teleſcope, the *Lumen Boreale*, and the electric flaſh, was ſuggeſted to the writer of this hiſtorical ſketch by the phaenomena themſelves, and not by the paſſages juſt quoted, of which he had not the leaſt knowlege till after he had formed his opinion on the ſubject, but which indeed he was glad to have fortified by ſuch good authorities.

It is with diffidence he mentions another teleſcopic appearance, as it may perhaps be only an optical illuſion. It ſeemed perceptible that the nucleus of the comet had a very rapid motion on its own axis, and that there was a conſtant emiſſion of electric ſparks, or flaſhes from it at all points, which were inſtantly impelled with great violence towards the tail.

It has not as yet been poſitively determined by aſtronomers, whether this preſent comet is one of thoſe that have ever viſited us before, at leaſt ſince comets have been ſo narrowly watched. But it is hoped that the worthy ſucceſſor of the great Flamſtead, Halley and Bradley, by whom this comet will doubtleſs have been well obſerved, will clear up this point, when he ſhall have had ſufficient leiſure for computing its elements, and comparing them with thoſe of other comets already computed. Out of above 50 comets whoſe elements are known, the returns of no more than 5 or 6 have been foretold: which may be ſeen in the following table.

Former Appearances.	Years Period	Return expec. about	Orbits by whom calculated.
———— 1532 : 1661	126	1789	Halley.
1531 : 1607 : 1682 : 1759	76	1834	Halley.
———— 1264 : 1556	292	1848	Halley.
1165 : 1338 : 1512 : 1686	174	1860	Pingre.
———— 1593 : 1762	169	1931	La Caille Pingre.
44 Ant. C. 531 : 1106 : 1680	575	2255	Newton, Halley.

The comets that are firſt expected, are placed firſt in the table. But it may be neceſſary to give the authorities upon which this table is conſtructed.

The comet of 1661 obſerved by Hevelius, and that of 1532 by Appianus, (though there is a conſiderable difference in the places of their perihelions, which may eaſily be attributed to the imperfect and unſkilful manner in which it was obſerved by Appianus) may be the ſame; and if Halley's conjecture be well founded, this comet will again appear in the year 1789 or 1790. (*Hiſt. des Mat. par Montucla.*)

Of the ſecond comet enough has already been ſaid, under the article 1759. Unleſs we here beſtow a few words upon the memory of M. Clairaut, who died May 3, 1765, aged but 52 years! in his elogium juſt publiſhed in the *Hiſt. de l'Arcad. des Sciences*, there is this paſſage, equally appertaining to that great geometrician and to this article.—' Thanks to his labours! the opinion that ' comets are planets, as ancient as the world it-' ſelf, is now no longer a conjecture, but is re-' garded as a thing fully demonſtrated.'

For the account of that which has the third place in the table, we are indebted to Mr. Dunthorne, (Phil. Tranſ. vol. 47.) who conjectures, that if the comet which appeared in 1264 be the ſame as that which was obſerved by Paul Fabricius and others in the year 1556, and whoſe orbit Dr. Halley has compared, its period will turn out as above.

Concerning the comet which occupies the fourth place in the table, we are indebted to M. Struick, who, in his ſequel to the deſcription of comets, (Amſt. 1753) ſuppoſes the comet of 1686 to be the ſame as that of 1512; " for one is upon record in 1338,—in 1165,— in 990,—in 817, —and in ſhort 870 years before, that is to ſay, at the diſtance of 5 times 174 years; 53 years before Chriſt."

Our authority for the fifth comet in the table, is given in the article for the year 1762.

And about the ſixth or great comet of 1680, which ſo alarmed the inhabitants of our globe in the laſt century, though the preſent is but little intereſted, ſo diſtant is the period: "Yet, if it return in 2255," ſays Maclaurin, "it will give a new luſtre and evidence to Sir Iſaac Newton's philoſophy, in that remote age. And ſhould the inconſtancy of human affairs, and the perpetual revolutions to which they are ſubject, occaſion any neglect of our philoſophy in the intervening ages; this comet will revive it, and fill every mouth again with this great man's name. For it is one of the good effects which theſe great periods, and diſtant depending obſervations promiſe, that they muſt contribute to preſerve the reliſh for learning from the revolutions to which it has been formerly ſubject. By them, diſtant ages are connected together, and perpetual matter is provided from time to time, for reviving the curioſity of mankind."

And though the returns of no more than ſix comets have, as yet, been calculated and foretold by aſtronomers; yet, from the elements laid down, every age will recognize ſome of thoſe which have viſited us before; till at length, the whole number of comets in the ſolar ſyſtem will be as well known to the inhabitants of the earth, as the five planets and their ſatellites.

One of them is already added to the number of revolving planets, which it is hoped will leſſen the popular dread at the appearance of the reſt. At this time of day it ſeems ſcarcely neceſſary to ſay, that the experience of many thouſand years aſſures us that the ſolar ſyſtem is ſo wiſely framed and regulated, as not to be ſubject to accidents from the approach of comets. But as two ſuch able aſtronomers and mathematicians, as Whiſton and Gregory, ſeem to encourage the belief that general ruin, or at leaſt dreadful calamities, may be the conſequence of a near approximation to the tail of a comet, and as their opinions have no other foundation than conjecture and ſpeculation, we may venture, without arrogance, to conjecture and ſpeculate in our turn.

And firſt, let us ſpeak to Mr. Whiſton, who having (as has been mentioned in M. Maupertuis's Letter) with great learning and labour traced the great comet of 1680 by different periods, of 575 years, up to the univerſal deluge, endeavours to account for that terrible cataſtrophe, by the water that was brought into our atmoſphere by the tail of that comet. We will, for a moment, grant it poſſible for comets to quit the regular and ſtated courſe originally aſſigned them, and likewiſe, for the ſake of Mr. Whiſton's hypotheſis, ſuppoſe the tails of comets to be compoſed of aqueous vapours (an opinion, which, with Dr. Hamilton's aſſiſtance, we have endeavoured to confute) yet ſtill a powerful objection to the deluge having been occaſioned by the tail of the comet remains; namely, that vapours at ſuch a great diſtance from the earth would be ſo rarified, that notwithſtanding the comet of 1744 had a tail 40 degrees long, the preſent comet one of 42, and that of 1680 one of 70 degrees in length—if all the waters of this enormous tail were compreſſed into the denſity of thoſe vapours in our atmoſphere, which, coaleſcing, compoſe rain; they would be inſufficient for the purpoſe of drowning the earth: ſince a ſingle drop of water, rarified into vapour in our groſs atmoſphere, occupies 14,000 times the ſpace it did, when condenſed into water; and vapours ſufficiently rarified to ſwim in Æther, though they formed a volume equal to the orbit of the Earth, would not furniſh water enough to overwhelm it, which Sir Iſaac Newton has demonſtrated by proving that a cubic foot of water, at the diſtance of a ſemidiameter of the earth, would be ſo rarified as to occupy a ſpace equal to the orbit of Saturn.

It has been ſaid above by M. Maupertuis, that Dr. Gregory has eſtabliſhed the comets again in all their terrors. In ſuppoſing ſuch dreadful conſequences from a comet's tail being in contact with our earth, &c.—if the learned Doctor were ſtill living, one would be inclined to aſk him, why the poor old ladies are ſtill to be frightened with comets? Do they either portend or occaſion any thing worſe than always ſubſiſts? When was the world without its plague, peſtilence and famine; battle, murder and ſudden death? Though comets are doubtleſs placed in the heavens for ſome wiſe purpoſe, wholly inſcrutable to us, yet, for any thing we have hitherto diſcovered, they have no more influence or effect upon our globe, or its inhabitants, than a *will-o'-th'-wiſp* or *ignis fatuus*;—and yet every thing that is miſchievous or diſagreeable is placed to the account of the poor comet.—If it rains, "it is the comet;"—if the weather be hot, "it is the comet;"—if it be cold, it is the ſame.—Pray let us be a little equitable, and allow that ſuch things, as abundant rain, intemperate heat, and intenſe cold, have happened in this climate before now, without the agency of a comet; unleſs by ſome comet that has appeared in diſguiſe, like Mr. Bayes's army.

"The war in the weſtern parts of Europe, which continued from the year 1688 to 1697, has been the moſt obſtinate and deſtructive of any recorded in hiſtory; and yet no comet has appeared, either immediately before or after; but on the contrary, one has been ſeen in September, 1698, when Europe was already freed from this war, and was on the point of eſtabliſhing a laſting peace between the Chriſtians and Turks.—A comet therefore has appeared between two treaties of peace, which have put an end to the ravages of war, in all parts of Europe, and greatly changed the general ſituation of affairs for the better: A comet which reſtores thoſe happy times, in which the temple of Janus is ſhut." (*Penſees div. ſur la comete de* 1680).

But ſuperſtitious people love to be frightened, and will be as angry with any one who endeavours to reaſon them out of their fears, as the inhabitants of *Neuf Chatel* were lately with one of their paſtors, who, though in other reſpects an orthodox and devout Chriſtian, yet could not reconcile to his belief the *eternity of hell torments.*—He would allow them to laſt a hundred thouſand years, with all his heart,—but that would not ſatisfy his flock,—they proſecuted, perſecuted, and pelted him. When the king of Pruſſia, their ſovereign, hearing of it, and moreover that the miniſter was a worthy, well-meaning man, ordered them to deſiſt, and ſuffer him to reſume his function. But this enraged them ten times more,—they ſurrounded the good man's houſe, and would certainly have ſent him to the other world, to enquire into the true ſtate of departed ſouls, had he not with great difficulty made his eſcape; and, at length, their ſovereign, finding how fond they were of everlaſting damnation, out of his great goodneſs, condeſcended to let them be damned to all eternity.—" And I alſo, (ſays the author from whence this account is taken) conſent with all my heart, and much good may it do them." —*Lettre de M. Baudinet.*

It is amazing that ſuch as are always ready to denounce divine vengeance, and to preſage every ſpecies of calamity to the frighted inhabitants of the globe, upon the appearance of a comet;—that they, whoſe belief of the interpoſition of a *particular providence* upon every trivial occaſion, is ſo firm, never ſhould think of extending their faith to the belief of a *general providence*, which has ſecured the globe from contingent evils!

But there yet remain many who will have it that " a comet never appears without blood,"—and are ſure to be right in their conjectures. For if Europe ſhould enjoy a profound peace, they have only to look at Aſia; and if all be quiet there, they have ſtill the other two quarters of the globe to fly to, which will, doubtleſs, furniſh them not only with carnage enough, but alſo with every other kind of evil, both phyſical and moral, their hearts can wiſh, to confirm them in their opinion.

But thoſe who are unwilling to ſee God, but in vengeance and diſtruction, ſhould try to diſcover him in his goodneſs and protection from general calamity, by that wiſe order of his providence, ſo viſible in the wonderful and ſtupendous arrangement of the univerſe.

POSTSCRIPT.

LONDON, Oct. 25. 1769.

THE comet is returned.—It was ſeen laſt Monday, by ſeveral people, in different parts of the kingdom, at half an hour paſt ſix in the evening; weſt by ſouth, about nine degrees high. It appears with a *Coma*, but without a tail: perhaps the tail may be viſible in a clearer air. The comet is at preſent more diſtant from us than the Sun, and its tail oblique to us, whereas, in its former appearance, it was perpendicular.

THE END.

www.ingramcontent.com/pod-product-compliance
Lightning Source LLC
Chambersburg PA
CBHW020801020726
47495CB00008B/2537
* 9 7 8 3 7 4 4 7 6 2 4 2 7 *